Outline Studies in Ecology

Editors

George M. Dunnet
Regius Professor of Natural History,
University of Aberdeen

Charles H. Gimingham
Professor of Botany,
University of Aberdeen

Editors' Foreword

Both in its theoretical and applied aspects, ecology is developing rapidly. This is partly because it offers a relatively new and fresh approach to biological enquiry; it also stems from the revolution in public attitudes towards the quality of the human environment and the conservation of nature. There are today more professional ecologists than ever before, and the number of students seeking courses in ecology remains high. In schools as well as universities the teaching of ecology is now widely accepted as an essential component of biological education, but it is only within the past quarter of a century that this has come about. In the same period, the journals devoted to publication of ecological research have expanded in number and size, and books on aspects of ecology appear in ever-increasing numbers.

These are indications of a healthy and vigorous condition, which is satisfactory not only in regard to the progress of biological science but also because of the vital importance of ecological understanding to the well-being of man. However, such rapid advances bring their problems. The subject develops so rapidly in scope, depth and relevance that text-books, or parts of them, soon become out-of-date or inappropriate for particular courses. The very width of the front across which the ecological approach is being applied to biological and environmental questions introduces difficulties: every teacher handles his subject in a different way and no two courses are identical in content.

This diversity, though stimulating and profitable, has the effect that no single text-book is likely to satisfy fully the needs of the student attending a course in ecology. Very often extracts from a wide range of books must be consulted, and while this may do no harm it is time-consuming and expensive. The present series has been designed to offer quite a large number of relatively small booklets, each on a restricted topic of fundamental importance which is likely to constitute a self-contained component of more comprehensive courses. A selection can then be made, at reasonable cost, of texts appropriate to particular courses or the interests of the reader. Each is written by an acknowledged expert in the subject, and is intended to offer an up-to-date, concise summary which will be of value to those engaged in teaching, research or applied ecology as well as to students.

Community Structure and the Niche

Paul S. Giller

Department of Zoology,
University College, Cork,
National University of Ireland

London   New York
Chapman and Hall

First published in 1984 by
Chapman and Hall Ltd
11 New Fetter Lane
London EC4P 4EE
Published in the USA by
Chapman and Hall
733 Third Avenue
New York NY 10017
© *1984 Paul S. Giller*

Printed in Great Britain by
J. W. Arrowsmith Ltd, Bristol

ISBN 0 412 25110 8

British Library Cataloguing in Publication Data

Giller, Paul S.
  Community structure and the niche.—
  (Outline studies in ecology)
  1. Biotic communities
  I. Title   II. Series
  574.5$'$247   QH541

  ISBN 0-412-25110-8

Library of Congress Cataloging in Publication Data

Giller, Paul S.
  Community structure and the niche.

  Bibliography: p.
  Includes index.
  1. Biotic Communities.   2. Niche (Ecology)   I. Title.
  QH541.G54   1984   574.5$'$247   84-4988
  ISBN 0-412-25110-8 (pbk.)

Contents

Preface

During the past two decades, there has been a gradual change of emphasis in ecological studies directed at unravelling the complexity of natural communities. Initially, the population approach was used, where interest lay in the way individual populations change and in the identification of factors affecting these changes. A good understanding of the dynamics of single populations is now emerging, but this has not been a very fruitful approach at the community level. In the natural world, few species can be treated as isolated populations, as most single species are the interacting parts of multispecies systems. This has led to a community approach, involving the study of interrelationships between species within communities and investigation of the actual organization of natural communities as a whole. The formalization of a number of new concepts and ideas has evolved from this approach, including niche theory, resource allocation, guild structure, limiting similarity, niche width and overlap etc., which, until fairly recently, have been examined mainly from a theoretical point of view. However, a wealth of field data is gradually being added to the literature, especially from the general areas of island biogeography and resource partitioning amongst closely related species.

Community structure embodies patterns of resource allocation and spatial and temporal abundance of species of the community, as well as community level properties such as trophic levels, succession, nutrient cycling etc. It would be difficult to approach all aspects of this complex and wide-ranging concept in a book of this size. However, by concentrating on two important indices of community organization, namely the number of species and their relative abundance, one can begin to explore the design and functioning of natural communities and also begin to identify the patterns and ground rules of their structure.

An understanding of how communities function has practical implications in such diverse areas as land and water management, management of crop pests, design of conservation regions, controls of vectors of tropical disease and management of world fish stocks. There is thus a clear need to provide answers to questions about the nature of the structure and dynamics of natural communities.

This book is intended to give an introduction to the current theories and ideas on community structure and to provide an opening into the vast and detailed literature now available. The study of community ecology is in a state of flux, and will continue to be so until sufficient field data are available to test thoroughly the current theories and indicate in which direction new theories should go. There is thus huge scope for further work in this rapidly expanding field of ecology.

Chapter 1

Introduction
and definitions

1.1 The community

At its simplest, the term community describes a group of
species populations occurring together, as in a pond or
woodland. However, many workers will refer to communities
of birds, insects or plants for example, which causes confusion
over the scale and true ecological meaning of the community.
The term assemblage is a more appropriate description for
such a group of similar species populations occurring together (i.e. an assemblage of birds, insects or plants). A
community of organisms should be viewed more as an
organized whole, and any definition should encompass
interactions among constitutent populations, i.e. an association of interacting populations of all trophic levels occurring
in a given habitat [1]. Species do adapt to the presence of
other species, so, just as populations have properties over
and above those of the individuals comprising them, the community is more than the sum of the individual populations and
their interactions [2]. Whittaker's definition [3] is the most
precise to date, describing a community as a combination
of plant, animal, and bacterial populations, interacting with
one another within an environment, thus forming a dis-
tinctive living system with its own composition, structure,
environmental relations, development and function. Despite
this precision, it is difficult to say what a natural community
is and how one recognizes it, so the concept of a community
is often an abstraction. Communities are, in reality, open,
generally intergrading continuously along environmental
gradients rather than forming clearly separated zones as

envisaged by early thinkers [2, 4]. Similar difficulties in identification have been faced by population biologists. Sometimes environmental heterogeneity and topographical barriers delimit a natural population; sometimes judgement, arbitrary selection or experimental demands are applied. The same criteria are used by the community ecologist. For example, some theoreticians simply specify an arbitrary set of species [5]. Another technique delimits communities objectively, using what is known as a species—area curve. By recording cumulative numbers of species in an expanding area, a characteristic curve results (Fig. 1.1). The minimal area that includes the community's representative species combination is given where the curve reaches its asymptote. Lake and woodland communities are somewhat easier to delimit, although one often arbitrarily considers only a part of such systems. Despite difficulties of definition, the study of the community is an important step in our study of the natural world as a whole.

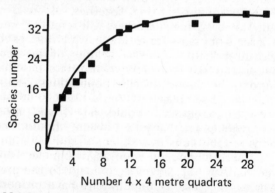

Fig. 1.1   Mainland species—area curve from a forest understory plant assemblage in North Carolina. (After McNaughton and Wolf [6]).

1.2   Community structure

It is generally believed that communities, as living systems of interacting species populations, are organized in some way, and that the role of the community ecologist is to unravel and explain that organization. One theme of this book is to identify and discuss the patterns which do seem to confer some degree of organization on to communities.

1.2.1   *Levels of study*

A possible method of investigating community organization

is at the individualistic level, where the behaviour and population dynamics of individual species are examined in terms of interactions between and within the populations. Such a method, originated for population studies of single species or species pairs, is difficult to extend to multispecies situations [7], so an alternative, holistic, approach tends to be used, focusing on the overall aspects of community structure. This type of investigation is helped by the concept of a guild; an assemblage of species utilizing a particular resource or group of resources in a functionally similar manner [8]. Members of such guilds interact strongly with one another and weakly with the remainder of the community. One could thus speak of an insectivorous bird guild or a habitat guild of lizards. This level of study is important, as guilds presumably represent the arenas of most intense interspecies interactions [9].

1.2.2 A definition

Plant and animal ecologists may appear to differ in their interpretation of the term community structure, but fundamentally, they both refer to the same phenomenon. Community structure embodies (a) all the various ways individual members of communities relate to and interact with one another (i.e. patterns of resource allocation and spatial and temporal abundance of species of the community); (b) the community level properties arising from these relations (such as trophic levels, succession, rates and efficiencies of energy fixation and flow, nutrient cycling etc.). It would be difficult to approach all aspects of this complex and wide-ranging concept in a book of this size. However one can examine the structure of communities by concentrating on two important indices of community organization, namely the number of species and their relative abundance [10]. Using these indices one can then try to answer the following questions:

(i)   How do species fit together to form a community?
(ii) What determines the numbers of species making up different communities?
(iii) How might the interactions between species populations set an upper limit to this number?
(iv) What are the implications of differences in relative abundances of species in a community?

With this information, the ecologist may then investigate

the similarity of community patterns from different geographical areas, validate the structured nature of natural communities, and begin to identify some of the 'ground rules' of this organization. In simple terms, ecologists observe that biotic communities differ dramatically in the numbers of plant and animal species they support (i.e. species richness), and that, given this species richness, communities show differences in the relative abundance of constituent species. We are eager to discover why!

1.3 Species diversity

In addition to simple species number indices, species diversity is often used as a more representative measure of community richness, as it incorporates both species number and relative abundance. The choice of index, from the bewildering variety available, depends on such factors as the difficulty in appraisal of species abundance and success in sampling and identifying all species present. The derivation, theory and use of such diversity indices can be found in several reviews [11, 12]. For many purposes, the number of species present is the simplest and most useful measure of local or regional diversity.

1.4 Trends in species richness

1.4.1 Latitudinal gradients

One method of estimating the number of species occurring within different regions is to partition maps of large land areas into equal sized quadrats, on which range maps of individual species are superimposed [13, 14]. These and other studies have revealed the well-known latitudinal gradients of species richness, where, in most groups of organisms, the number of species increases markedly towards the equator. An obvious example is a comparison of the variety of trees in most tropical rain forests with the solid stands of timber in boreal regions. Nesting birds show a typical latitudinal gradient (Fig. 1.2) and Fischer describes similar gradients for ants, corals, tunicates, amphipods, nudibranchs and gastropod molluscs [15]. More recent examples include American insectivorous birds [14, 16], lizards (Fig. 1.3) and Australian endemic *Drosophila* [17]. One of the drawbacks to such studies lies in the fact that the number of habitats in a given quadrat or area relates to topographical relief. Another is that there is a greater diversity of habitats in low latitudes (e.g. ranging from tropical to boreal with altitude) than in higher

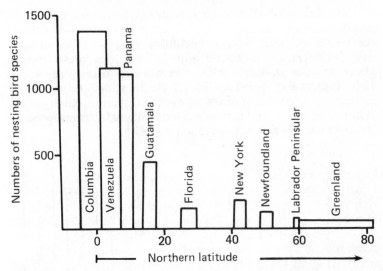

Fig. 1.2 Latitudinal diversity gradient in nesting birds. (From Fischer [15].)

latitudes (which progressively lose this range), so it is not surprising that on this gross scale, more species are found in the tropics. Nevertheless, a comparison of similar habitats, for example at high altitude, still reveals greater species richness in the tropics [9].

Fig. 1.3 Geographic patterns in species densities of lizards in the United States. (After Schall and Pianka [14].)

1.4.2   Habitat gradients

Smaller scale studies compare species richness across many different habitats within latitudinal belts. These usually reveal differences between adjacent habitats, even though there are no physical barriers preventing species from one habitat invading another (Fig. 1.4). In addition, consistent trends in species numbers involving altitude, topographic relief, island size and location, peninsular effects and proximity to oceans have been documented [15].

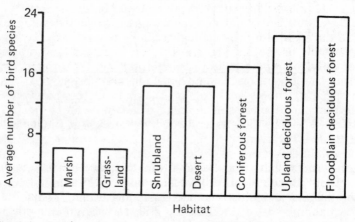

Fig. 1.4   Species richness of birds in representative temperate zone habitats. (After Tramer, E. (1969), *Ecology*, **50**, 927—29.)

1.4.3   Exceptions to the rule

Latitudinal trends are not universal. The gradients are not shown by burrowing marine invertebrate groups like Ophiuroids and Holothuroids which show little diversity anywhere. Similarly, the prosobranch mollusc family Naticidae, a soft bottom dweller, shows no trends, whereas the epifaunal prosobranchs show good latitudinal gradients [15]. Latitudinal trends are virtually non-existent among Australian vertebrate taxa [14], and are often not very clear in plant assemblages apart from forests [18]. The prevalent trend is also occasionally reversed, often by small specialized taxonomic groups. For example, sandpipers and plovers are more diverse in the Arctic [10], a greater diversity of breeding birds is found at higher latitudes in Eastern deciduous forests of the USA [16], and marsupials appear to be more diverse in temperate regions than the tropics [14]. Red algae and kelps also show greater diversity in temperate regions [15]. These exceptions to the rule are, nevertheless, worthy of further investigation. Elucidation of their controlling factors is as important to our understanding of community structure as is success in explaining the general trends described earlier.

1.5 The problem restated

These repeated patterns in species richness suggest that general explanations may be possible. If we accept the premise that a community is a structured assemblage of organisms, then it is the interactions between these organisms that should provide the structure. The ecological niche is a reflection of the organism's or species' place in the community, incorporating not only tolerances to physical factors, but also interactions with other organisms. The obvious question posed by patterns in species richness is 'What are the main factors controlling the number of species in an area?' This can be more precisely stated as 'What factors control the number of niches in a given area?', given that individual species utilize different niches. Terms such as species packing, niche overlap and niche width have arisen as a direct result of theoretical attempts at solving this problem, and have become incorporated into an almost separate discipline, Niche Theory.

Chapter 2

Niche theory

The concept of the niche pervades all of ecology, yet it has become somewhat confused through popularization and attempts to make objective sense out of an originally subjective idea (e.g. use of a systems approach [19] or mathematical integration of related concepts [5]). It arose as an attempt to describe the total role of a species in a community, defining all the bonds between populations, community and the ecosystem. As such, the niche relates the concepts of the population and community, describing how ecological objects fit together to form enduring and functioning wholes, and enabling us to see how very different communities may resemble each other in the essentials of their organization. It is this role that has led to its rise in importance over the last 30 years.

2.1 Development of the niche concept

This has been discussed in detail elsewhere [5, 20], but a brief review would be useful.

Grinnell [21] introduced the term niche as a habitat concept, defining the ultimate distributional unit of a species. He implied that niches of species do not overlap, and thus identified the potential nature of a species' distribution in the absence of interactions with other species. Elton's independent definition of the niche encompassed mainly a functional concept, describing an organism's place in the biotic environment in terms of its relations to food and enemies [22]. He was, in effect, referring to a species' actual rather than potential place in nature.

2.1.1 The competitive exclusion principle

At the same time, the associated concept of competitive exclusion was developing. This suggests that two species with identical ecologies cannot survive simultaneously in the same place. This idea was implicit in Darwin's writings [23], and qualitatively stated by Grinnell, but apparently excited little interest at the time. However, between 1920 and 1940 mathematical demonstrations (Lotka—Volterra equations) and controlled laboratory experiments (the famous studies of Gause and Park), showed that competitive exclusion will often occur in the establishment of a two species population equilibrium. Since then, the principle of competitive exclusion, stated in the form 'complete competitors cannot coexist indefinitely', has become one of the central tenets of theoretical ecology [20]. In the present context, the corollary of this principle is important. If two species do coexist, then there should be some ecological difference between them, implying such species each have their own unique niche.

The Competitive Exclusion hypothesis could be considered of little scientific worth, as it is untestable (e.g. [9, 20]). However, it has been of immense value, both in the development of the niche concept and in prompting ecologists to answer such questions as: how do similar species coexist? how much difference between species allows coexistence? and how is competitive exclusion avoided? These questions have directed research in a way that provides a better understanding of community organization.

2.1.2 The multidimensional approach

The niche of the 1940s and 1950s took on the vague definition of an organisms 'profession' within the community (e.g. [24]), but thereafter achieved a formal and potentially quantitative definition through the work of Hutchinson [25]. He considered the niche to be defined by the total range of environmental variables to which a species must be adapted (physical, chemical and biotic), and under which a species population lives and replaces itself indefinitely. Ideally, every pertinent environmental variable can be considered as a gradient along which the species has an activity or tolerance range. An example is light intensity, which suffers a logarithmic extinction from the forest canopy downwards as light is intercepted by plants [11]. The species evolve to relate themselves to this gradient, each adapting to a different range of light intensities. In some boreal forests this leads to an ideal, size related structure containing 5—7 species, including canopy and smaller trees, tall and low shrubs and herbs, and a ground level moss [11].

Each environmental gradient can be thought of as a dimension in space. If there are *n* pertinent dimensions, the niche can be described in terms of an *n*-dimensional space, or hypervolume. Potentially, this can be built up one dimension at a time. Fig. 2.1 shows a species response to one environmental gradient, where some measure of fitness is normally distributed about a preferred point on the gradient. Simultaneous response patterns to two and three resources can be depicted graphically (Fig. 2.2) and this procedure can be extended to any number of axes using *n*-dimensional geometry [9], producing a very complex hypervolume representing the responses of the species population to all environmental factors. (This assumes all relevant variables are included and are independent of each other.) Hutchinson further defines two states of a species niche. The *fundamental niche* describes the entire set of optimum conditions

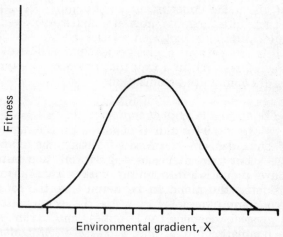

Fig. 2.1   A species response to a single environmental gradient. Measures of fitness include reproductive success, population size and survivorship.

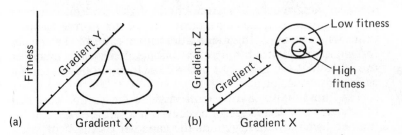

Fig. 2.2 Simultaneous species response to two environmental gradients (a), and to three environmental gradients incorporating species fitness (b). (After Pianka [9].)

which a species can occupy in the absence of enemies. The *realized niche* is the actual set of conditions in which the species normally exists. This is less than or equal to the fundamental niche.

This multidimensional approach provides a means of conceiving how species relate to one another and has thus enhanced our interpretation of community organization. In crude terms, one can think of total niche space of a habitat as an '*n*-sided' box, into which the niches of all species of the community fit, just like '*n*-sided' balls. If niches are always discrete (i.e. no overlap of fundamental niches), then the species richness of a community depends on the total amount of niche space (a habitat variable) and the average size of each niche (a species variable). Wide niches should lead to a lower species richness than narrow ones for a given variety of resources.

2.1.3 *The niche as a resource utilization spectrum*

Niche width (the size of a niche) is of central importance in niche theory, as an understanding of its controlling forces will lead to a greater understanding of the causes of species diversity. Under the multidimensional approach, niche width is defined as the sum total of the variety of different resources exploited by a species population. Its measurement at this level requires a description of all relevant parameters, and a continual estimation of simultaneous proportional utilization of resources, which is clearly an impossible task. Thus the ecological niche has become identified increasingly with the distribution of species activity (resource utilization spectra) along just one or a few of the most important (or most easily measured) niche dimensions [5, 26—28]. The niche of each species is then defined by a utilization function (the distribution of species' activity) along a resource gradient (Fig. 2.1). Provided the niche dimensions examined are truly independent, overall multidimensional utilization may be thought of as a product of the individual, unidimensional utilization functions.

The most important characteristics of a niche described in this way are the height (maximum rate of resource utilization or level of activity) and the breadth of the utilization curve, the latter providing some limited but attainable measure of niche width.

2.2 Niche width

Two separate components combine to form the niche width of a species [26]. The within-phenotype component (WPC) describes the level of variation in resource use by individuals, and the between-phenotype component (BPC) describes variation amongst individuals of the species population. Total niche width (B) is given by WPC + BPC. If B is 100% BPC, the species will be polymorphic with specialists, whereas if B is 100% WPC, the species will be monomorphic with generalists. Obviously real populations will lie somewhere in between (Fig. 2.3).

Two basic procedures for the measurement of resource utilization have been identified [29]. The first involves a simple description of the species utilization of a continuous resource in terms of the mean (d) and width (w, as 1 standard deviation) of the utilization curve on the resource gradient. A large w indicates a wide niche. Adjustments are possible when w is different on either side of the mean [29]. This

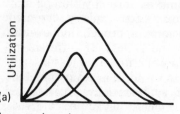
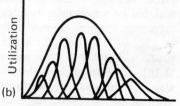

Fig. 2.3   Idealized representation of populations differing in the components of niche width. (a) High within-phenotype component; (b) high between-phenotype component. (After Pianka [9].)

measure of niche width is commonly used to assess resource utilization on the basis of morphological variation in a trait related to resource use, e.g. feeding structures [26, 30]. The second method does not require resources to be ordered along a continuum, but is based on proportional utilization of various resource states (e.g. prey species). Broad niche species tend to use resources in proportion to their availability, whereas narrow niche species will tend to concentrate on items in only some resource states. A number of different indices of niche width have been formulated (e.g. [31, 32]) and a comparison of those most commonly used is provided in a recent review [33].

There are certain limitations involved in the use of such indices. For example, because the width of niches only relates to the available width of the resource spectrum, one cannot readily compare the width of one species' niche with that of another using a different and unequal resource spectrum. Also, the accuracy of the measure of niche width depends on how objectively the ecologist defines the available resources [33]. Niches are not smooth curves along a few, simple, dimensions of the environment, and our ability to measure niches may fall far short of reality.

2.3 Niche overlap

Most organisms do not inhabit their potential, fundamental niche, but, due to interactions with other organisms, occupy a reduced, realized niche. The major interactions are normally considered to be predation and competition, and the latter has become involved in niche theory through the concept of niche overlap. Rather than niches in a community appearing as discrete, non-interacting units, species tend to share parts of each others fundamental niches, resulting in simultaneous demands upon some resource by two or more species populations. In Hutchinson's terminology, the niche hypervolumes of species include parts of others, thus overlap.

2.3.1 Possible outcome of niche overlap

If the overlap is very small, or the resources superabundant, then these species can coexist in essentially separate and almost fundamental niches. If niches overlap to a greater extent and resource availability in the overlap zone cannot meet demand, then the abundance of the less-efficient species will be limited by interactions with the more efficient. Ultimately, competitive exclusion may occur in the overlapping parts of any two niches. Making this an assumption, one can consider the hypothetical outcome of different degrees of niche overlap between two species [9].

(i)   Under the improbable situation of the two fundamental niches being identical, the competitively superior species would totally exclude the other.

(ii) One fundamental niche might be totally included within a second, larger one. Here, an inferior included species would be eliminated, but a superior included species would eliminate the other species from the contested space (Fig. 2.4(a)).

(iii) With partial overlap of fundamental niches, the competitively superior species occupies the shared niche space, and each species has an exclusive, uncontested refuge (Fig. 2.4(b)). Coexistence is thus theoretically possible, but will depend on the amount of overlap which can be tolerated by the inferior species.

(iv) Niches may abut against each other (Fig. 2.4(c)). No direct competitive exclusion can occur, but such niche relations might reflect the avoidance of competition.

(v)   Niches are entirely disjunct, so both species occupy their fundamental niche (Fig. 2.4(d)).

For example, most forests contain many more than the 5—7 plant species described in an earlier example (Section 2.1.2). The additional species will also utilize the gradient

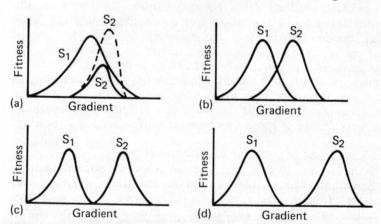

Fig. 2.4   Possible niche relationships between two species on a single environmental gradient. (a) An included niche; (b) overlapping niches; (c) abutting niches; (d) disjunct niches. (After Pianka [9].)

of light intensity, fitting in between the centres of population of other species. On the basis of the above exercise, the inclusion of extra species along the gradient should decrease the niche width of species already present, and lead to the packing of increasing numbers of plant species along the same gradient of light intensity.

2.3.2   *Measurements of niche overlap*

It has often been suggested that the key to understanding species interactions in a community is to measure the degree that niches of two species overlap, rather than trying to describe the niches of all species [10]. Such overlap is usually measured in terms of utilization data of resources such as food and microhabitat (a combination of the important and easy to measure factors). Niche overlap is thus described as overlap of utilization between two adjacent species on a resource gradient.

The simplest measurements are based on separation of resource utilization functions, and overlap is described by the following resource separation ratio [29]

$$\rho_{ij} = d_{ij}/w_{ij}$$

where d_{ij} is the difference between the means of resource utilization by species i and j, and w_{ij} is the common width of the utilization curve (1 standard deviation), given by

$$w_{ij} = (w_i^2 + w_j^2/2)^{\frac{1}{2}}$$

If ρ_{ij} is less than 3, there should, theoretically, be some interaction between the species. There should also be some minimum separation value below which competitive exclusion operates (see Chapter 4).

More complex measurements are based on a variety of methods including percentage similarity, chi-square goodness of fit and information theory. At least eight different indices are in current use and have been examined critically elsewhere [34—36]. A picture of niche overlap between all members of a guild or community can be built up using a niche overlap resource matrix [9]. An *m* by *n* matrix is constructed, indicating the amount of each of *m* resource states utilized by each of *n* different species, and from this an *n* by *n* matrix of overlap between all species pairs can be generated.

One can also assess the combined overlap along two or more resource dimensions to obtain some measure of total overlap between species. For independent resources, the product of individual overlap measures is used, but if the resources are dependent ones, summation of overlap values is necessary [35]. No method has yet been devised to allow for various degrees of independence between resource dimensions.

In the measurement of niche overlap, allowances must be made for the fact that continuous resource dimensions do not provide equal ecological opportunities along their entire length. For example, small prey are likely to be more abundant than large prey. It is also possible for significant differences in resource utilization to occur between different weight, size or age classes of a species (e.g. [37]). Such intraspecific differences must be accounted for when niche overlap comparisons are made with other species. (See also Section 4.2.)

2.3.3 *Niche overlap and competition*

Niche overlap values are frequently equated with the competition coefficient (α) of the classical interspecific competition equations of Lotka and Volterra [5, 10, 36]. However, such comparisons are fraught with biological difficulties [29] and the actual relationship between niche overlap and competition is not clear.

Mere overlap in resource use does not necessarily lead to competition, as assumed in Fig. 2.4. Likewise, the intensity of competition need bear no relation to the degree of niche overlap [37].

The more abundant the resource, the less likely it is that competition will result from its common use, and no competition is expected between species sharing an unlimited resource in some habitats (oxygen is an obvious example in most terrestrial systems). Thus the ratio of demand to supply, or the degree of saturation, of the environmental resource is of vital importance in the relationship between niche overlap and competition. Few authors, however, have actually incorporated resource availability into measures of niche overlap and competition [38]. In addition, overlap on one resource gradient may indicate diversification in other ways. (See Chapter 4.)

Finally, an inverse relationship between competition and niche overlap has been suggested [9], predicting that maximum tolerable overlap should be lower in intensely competitive situations than in environments with lower demand/supply ratios. (This Niche Overlap hypothesis will be discussed in more detail in Chapter 3.)

2.4 Diffuse competition

Consideration of niche overlap has led to another aspect of niche theory, diffuse competition. A species niche will usually only overlap with a limited number of adjacent niches on one resource gradient, but the potential number of neighbours increases as one examines more and more environmental dimensions simultaneously. Therefore, although pairwise niche overlap may be small, the cumulative effect of this diffuse competition can severely reduce the size of the realized niche, even to the point where it is too small to support a viable population (Fig. 2.5). A species can thus be 'squeezed out' by a group of other species.

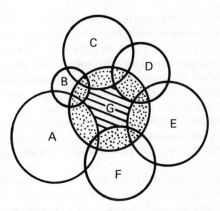

Fig. 2.5 Diffuse competition can reduce the fundamental niche of species G (stippled plus cross-hatched area) to its realized niche (cross-hatched area only).

The number of potential niches in a community can now be considered as a function of the degree to which development of the community leads to separation of partially overlapping niches under a given environmental regime [39]. In terms of the simple 'ball-and-box' model, each ball's volume can be decreased by squashing more balls into the box. The resilience or flexibility of the balls will control how many of them can then be packed into the available space.

2.5   Niche dynamics

Our ability to visualize and study an organism's niche, and its
interactions with those of other organisms, is certainly cur-
tailed by its ability to change both in time and from place
to place as the physical and biotic environment varies (i.e.
the niche can change position within the total niche space).

Temporal changes can be short term, i.e. within the life
of an individual or over a few generations (ecological time
scale). This is especially true of organisms undergoing some
form of metamorphosis during development, such as holo-
metabolous insects, planktonic and sessile crustacea, most
aquatic insects, amphibians etc. Such organisms have disjunct
niches at different times in their life histories [9]. Other
organisms such as hemimetabolous insects, and predators
utilizing different prey sizes as they grow, may show a more
gradually and continuously changing niche. Niche changes over
evolutionary time have undoubtedly taken place, where the
opening up of new adaptive zones has allowed scope for their
exploitation through evolution and adaptive radition.

On a smaller scale, the size or width of a realized niche is
likely to change through the responses of a species or its
competitors to changes in resource levels or to activities of
the resources themselves. Such changes are predicted by
optimal foraging theory which is based on the idea that individual consumers should maximize individual fitness
(usually through maximization of net energy gain) through
their foraging behaviour (e.g. [40]). Finally, the community
niche space is likely to shift in daily and seasonal rhythms,
so the inter-relationships of each species in the community
must also continually change. In terms of the simple model
of the community, not only can the 'balls' alter in shape, size
or position within the 'box', but the volume and shape of
the 'box' can also change. It is these dynamic properties that
make the niche so difficult to measure, so that at best one
can obtain fragmentary measures which indicate the relative
degrees of niche width and overlap within communities, and
use these to help unravel the complexities of community
organization.

2.6   The niche — a property of the species or the community?

Niches and organisms form complementary pairs, and one
view, noting the role that the organism itself plays in creating
and defining the environment in which it lives, suggests that

niches are generated by the occupant [19]. This is acceptable in part, when one considers the construction of artifacts such as nests, termite hills and beaver dams, as these modify the environment and contribute part but not all of the niche of an organism. An alternative view is that the niche is really a property of the community, and has no meaning except in the community context [39]. This implies that niches are generated by abiotic and biotic ecosystem components, are thus pre-existing, and are filled through species adaptation over a period of evolutionary change. One would therefore expect that communities in ecosystems characterized by similar environments should be of similar construction, and contain one or more essentially identical niches. The adaptations of populations filling such niches in these independently evolved communities should also be similar, even though the species may be totally unrelated. This is the phenomenon of ecological equivalence or convergent evolution, and its existence provides support for the community status of the niche.

The existence of habitat types that can be grouped together as biomes implies some form of community convergence. At a more specific level, succulent desert plants in the USA are cacti and these are almost identical to the plants of the spurge family (Euphorbiaceae) in South Africa [39]. Convergent evolutionary responses are also found in desert lizards. North American and Australian deserts each support a cryptically coloured, hornily armoured ant specialized species, a medium sized, lizard-eating species, and a long-legged species found in open spaces between plants. Similarly, Africa and Australia have convergent pairs, such as subterranean species [41]. Ecological equivalence is frequently documented in birds, where, for example, it is possible to match morphology and ecology of single species of Mediterranean birds on three continents [42], and similarly, nine species pairs of ecological equivalents have been found in Panama and Liberia (Fig. 2.6).

For two unrelated species to evolve to a point where they are almost identical, the niches to which they have adapted must also be almost identical. This is impossible if the niche is a property of the species population. This conclusion is further substantiated by species turnover on islands which show a stable species richness (see Chapter 7).

However, among larger guilds, or corresponding guilds of unequal species number, such exact matches appear to become obscure, and species replacement of two for three, three for

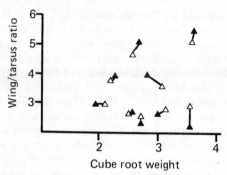

Fig. 2.6   Morphological similarities of ecological equivalent bird species in Panama (▲) and Liberia (△). (After Karr and James [43].)

five and other complex bases can occur [43]. This implies that in these situations the observed niches may be more a property of the species present.

Is the niche then a property of the species or the community? It seems that a certain ecological niche space is created by the physical and biotic components of an ecosystem, and this is a community property. The corresponding space in two similar saturated ecosystems can be divided amongst their species on a one to one basis, producing ecological equivalence, and at this level the niche appears to be a community property. Complex ratios of species replacement between two similar ecosystems may be due to historical factors, taxonomic barriers to convergence, or differences in the state of resources. These will influence the number and type of species present and the likelihood of equivalence. The observed niches in one or both communities may then be more a property of the constituent species.

2.7 Summary

The community can be thought of as a large, *n*-dimensional hyperspace, within which each species population evolves towards its own portion of the space. The position of the species and its response to factors of the community hyperspace defines its niche. Each species thus occupies a vaguely outlined, diffuse volume that differs from but perhaps overlaps with, those of other species in the community. The size and position of the niches are probably changing over both ecological and evolutionary time scales.

The full potential of the genotype of a species population

in all ecological matters has been termed its ecopotential [294]. However, the distribution and abundance of species is ultimately determined by tolerances to extremes of physical conditions (the fundamental niche), and species are usually further limited to some smaller range of habitats and population size by interactions with other organisms (the realized niche). Hence, if communities are organized by such interactions, then the manner and degree of organization will be reflected in differences between sizes and shapes of the realized and fundamental niche. Competition and predation are two major species interactions influencing these differences. In the following few chapters the evidence for these interactions and their effects on niche size, species packing, species richness and community organization will be examined.

Chapter 3

Competition and the niche; the effect on niche width

By definition, competition occurs when interaction between two or more individuals or populations adversely affects growth, survival, fitness and/or population size of each, typically when a common resource is in short supply. The interactions can be direct through interference, territoriality etc., or indirect through exploitation, involving joint use of some limited resource. Competition can occur between individuals of the same (intraspecific) or different (interspecific) species, and both types of interaction have important consequences in the community. Indeed, competition, and especially interspecific competition, is widely regarded as the principal mechanism determining ecological diversity [8, 44, 45]. The Competition hypothesis is as follows. Competition has adverse effects on all species utilizing a similar and limited resource at the same time and place (potentially leading to competitive exclusion of some species). It will thus be advantageous for a species population to obtain a degree of protection from competition with other species whenever possible. Natural selection should favour individuals in refuge areas of niche space, and so lead to reduced overlap in resource use and diversification of niches amongst the species populations. Competition thus influences the size of the realized niche, which in turn, is one factor influencing the species richness of a community.

3.1 Theoretical effects of competition on the species niche

On the basis of the preceding discussion, intraspecific and interspecific competition are likely to influence niche width in different ways. The following, simplified, theoretical analysis produces testable hypotheses. These can then be investigated in the field and laboratory to examine the actual role of competition as an organizing factor in the community.

3.1.1 Intraspecific competition

Where a normal utilization function is appropriate (Fig. 2.1), the resources are not being utilized to the same degree. Those individuals using marginal but less hotly contested points on the resource gradient will often have higher individual fitness than those using the heavily used, optimal resources [46]. The principle of equal opportunity, therefore, predicts that an individual's behaviour should lead to equilibration of the level of intraspecific competition through equalization of the demand to supply ratio along the resource continuum [28].

With an expanding population, the first individuals will use optimal resources, but as population density increases, the advantages to those individuals are offset by increasing intraspecific competition. This may favour deviant individuals that use the less-optimal, less-contested resources, leading to an increase in the variety of resources and habitats utilized by the species population as a whole (Fig. 3.1(a)). A decrease in the resource levels themselves will lead to the same predictable effects on niche width. Optimal foraging theory, based on energetics and an individual's fitness, predicts an expansion of diet and range of foraging areas utilized when resource levels or resource quality are reduced [40, 47, 48, 49].

Intraspecific competition thus tends to increase niche width, either through an increase in the within-phenotype component of the niche (behavioural or physiological flexibility), or through an increase in the between-phenotype component (differences between individuals). The changes in this latter component have been predicted more directly in the Niche Variation hypothesis [30]. This states that under conditions of relaxed interspecific competition (and presumably, therefore, increased intraspecific competition),

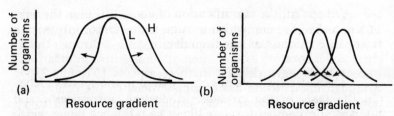

Fig. 3.1 Niche width changes under competition. (a) Intraspecific competition; L — Low density utilization curve, H — High density curve. (b) Interspecific competition. Arrows denote direction of change. (After O'Connor *et al.* [46].)

the increase in variety of resources available for specialization may permit greater phenotypic variation within a species population.

3.1.2 Interspecific competition

It is predicted that this restricts the range of the resource spectrum used by a species [27, 50]. Individuals of one species using marginal resources presumably cannot exploit them as efficiently as individuals of other species for which these resources are nearly optimal. The area of overlap between niches will be reduced, so niche width is decreased as the niches become more specialized (Fig. 3.1(b)). The population sizes of one or both of the species are also likely to decrease. The concept of diffuse competition is pertinent here.

Optimal foraging theory is less specific in its predictions. A more specialized diet is expected where a competing species selectively reduces levels of a particular food type or patch. On the other hand, an increase in the range of prey eaten (diet width) is predicted when the competitor reduces levels of food types or patches equally [40]. Pianka [27] formally proposed an inverse relationship between intensity of competition and degree of niche overlap between species pairs in his Niche Overlap hypothesis. When demand for resources is low relative to supply, potential competitors should tolerate a relatively high degree of overlap in resource use without experiencing critical levels of competition. When the demand to supply ratio is high, less overlap is tolerated. Thus, increased separation of niches is predicted as species diversity increases, with, presumably, a related decrease in the size of the realized niches of constituent species, as a result of interspecific competition.

3.2 Criteria for the identification of competition in the field
The potential for competition to be an important organizing factor in communities has been demonstrated through theoretical and laboratory examination of competitive interactions. Recent texts have dealt with these aspects [51, 52, 295]. Although competition has become crucial to the conceptual basis of much of community ecology, it has proved to be exceedingly difficult to study in natural communities. (This is presumably because competition is so potent that it is rarely possible to see the interactions that lead to exclusion.) To overcome this difficulty, a set of useful criteria have been drawn up to confirm whether or not competition is occurring [53]. These are: (a) the comparative distribution and/or relative abundance of two potentially competing species should be amenable to an explanation based on competition; (b) that the competing species are utilizing a common resource; (c) there should be evidence of intraspecific competition from the performance of a particular species in the field; (d) the results from the separate manipulation of the resource and the competing species should be predictable on the basis of the Competition hypothesis.

Very few studies have come near to fulfilling all these criteria, but in the following sections, I will illustrate the available evidence which demonstrates the impact and importance of competition on community structure.

3.3 The natural effects of intraspecific competition
These are easily seen amongst plants, where the tendency for large individuals to exert more influence on each other than do smaller plants sometimes leads to changes in the distribution pattern as the plants age. For example, as plant size of two desert shrubs *Ambrosia dumosa* and *Larrea tridentata* increases, dispersion patterns change from clumped to random to regular [54]. This is attributed to the phenomenon of self-thinning, caused by competition for limiting nutrients or water as root systems increasingly overlap. For these patterns to develop clearly, there must be little or no seedling establishment from year to year. Some sessile intertidal animals demonstrate similar intraspecific effects, where nearest neighbour distances increase as the sum of the sizes of neighbours increase [55]. Intraspecific competition is also inferred from 'buffer effects' in population studies,

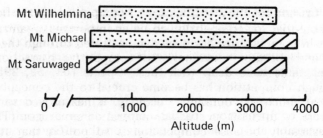

Fig. 3.2 Altitudinal ranges of honeyeaters on three New Guinea mountains showing competitive release of each bird species in the absence of the other. Hatched bar — *Ptiloprora peristriata*; stippled bar *P. guisei.* (After Diamond [57].)

where marginal habitats are used by territorial animals at high population densities [46].

3.3.1 Ecological or competitive release

Islands normally contain fewer species than comparable mainland systems, and thus the potential range of resources is increased. This has resulted in the widely documented phenomenon of competitive release. For example, island birds frequently show expansion in altitude range, habitats, diet, foraging areas and techniques [45], and increases in sexual dimorphism [56] compared with mainland populations of the same species. Mainland 'islands', such as isolated mountains provide similar examples (Fig. 3.2), and on mainland sites where close competitors are naturally absent, small mammals [58], birds [42, 59] and estuarine amphipods [60] have shown an expansion of habitat and/or diet ranges. Gorman provides more detailed examples [61].

Ultimately, competitive release of species can lead to adaptive radiation and convergent evolution. This is demonstrated by the *Geospiza* finches of the Galapagos islands, marsupial equivalents of placental dogs, cats, deer, mice etc. in Australia, and the radiation of weeds and bushes producing woody 'trees' on numerous islands [45].

3.3.2 Density compensation

If any site has a relatively incomplete biota (as on remote islands for example), then the existing species have access to additional resources normally utilized by the missing species, so species' densities are likely to increase due to increases in niche width on the food axis. The result is that two sites with differing species number may have similar total population density. Cody [42] has demonstrated a relationship between the density of bird pairs per acre (D) and species number (S) in species-saturated habitats, described by the equation:

$$D = 0.31S$$

If a species that is normally present in a habitat is missing, then no compensation, partial compensation or complete compensation are possible, depending on the efficiency with which the remaining species can utilize the vacated resources. Examples of all three occur in nature. Complete density compensation has been found for birds on Bermuda [42] and small lake islands in Sweden [62], and among small mammal assemblages [58]. Conversely, little or no compensation has been documented from bird assemblages on the satellite islands of New Guinea [57]. Other studies have shown partial compensation in birds [63], ants [64] and desert rodents [58, 64]. Density compensation appears to be most readily identifiable in exceptionally poor or exceptionally rich habitats.

The possibility of density compensation in plants has also been discussed [65].

3.3.3 Van Valen's niche variation hypothesis

Van Valen's studies [30] demonstrated that the variation in bill width of island birds exceeds that of mainland populations, and he concluded that there is a positive relationship between niche width and morphological or genetic variability. Data supporting this hypothesis also provide substantial evidence for the role of intraspecific competition on niche width, as an increase in morphological variation may provide a proximate cause of competitive release.

Direct support for the hypothesis has been found in studies on *Drosophila* species [67, 69] and foliage gleaning warblers [68]. The increase in sexual dimorphism of island birds described earlier, also represents an increase in both morphological variability and niche width under conditions of reduced competition. Indirect support for the hypothesis has been described from bat assemblages [66], ants [51] and *Anolis* lizard guilds [26].

The most often quoted evidence to counter this hypothesis is that some euryphagous birds (with a wide diet) are no more variable than stenophagous ones (with a narrow diet) [70]. However, euryphagous species can have either a large between-phenotype niche component, or a large within-phenotype component [71]. One would only expect an increase in variation within the population in the former case. In this study and in others which show ecological release but no complementary increase in morphological variation of the species (e.g. [63]), no account is taken of behavioural specializations or flexibility of the population which may form the basis of much competitive release.

3.4   The natural effects of interspecific competition
These effects have been studied in a number of ways. One of the most dramatic has involved attempts to demonstrate competitive exclusion in nature. Another has been to document changes in the niche of a species when faced with a competitor (niche shifts and character displacement). The changes in species composition and niche size when resource levels of a habitat alter can also provide evidence of the role of competition. Finally, the testing of the Niche Overlap hypothesis has provided another approach.

A review by Schoener [296] of over 150 experimental studies of interspecific competition in the field has revealed evidence of competition in 90% of experiments and amongst 76% of species involved (divided equally amongst aquatic and terrestrial plants and animals). Connell [297], in a similar review paper, presents lower figures, but the data base was considerably smaller and much harsher and more personal criteria were used in selection of 'suitable' studies. In both reviews not all the positive studies show competition amongst all species, at all places or for all of the time, but competition theory does not predict that it should [296, 298].

3.4.1   *Demonstration of competitive exclusion*
Three approaches have been used; (a) observations from natural 'experiments'; (b) manipulation of communities; (c) inference from comparative species distributions.

(a) Natural examples of competitive exclusion are most easily demonstrated in sessile organisms. For example, competition for space between colonies of different bryozoan species leads to overgrowth, which in turn usually leads to the demise of the undercolony. Three types of overgrowth can occur (Fig. 3.3), and clear overgrowth dominants can be found. Naturally changing environmental conditions can also lead to competitive exclusion. Population studies on fish have demonstrated the exclusion of rudd (*Scardinius erythroph-*

Fig. 3.3   Competitive exclusion in bryozoans. Arrows indicate direction of overgrowth. (After Jackson [72].)

thalmus) and perch (*Perca fluviatilus*) by roach (*Rutilus rutilus*) during the habitat changes involving the processes of lake succession [73]. The competitive superiority of roach during the young stages of the three species (when diets overlap) is thought to be the cause of such exclusion. The annual succession of marine and freshwater plankton species can be attributed to changing seasonal conditions which alter their relative competitive abilities.

Competitive exclusion of native forms by species introduced by man is a widespread phenomenon. The introduction of white fish *Coregonus* spp. in Sweden usually led to the virtual disappearance of the char *Salvelinus alpinus* in many char waters in the 19th and 20th centuries [50]. Extensive evidence of competitive exclusion in plant assemblages is given by Harper [75] and Tilman (295).

The spread of the grey squirrel, *Sciurus californensis*, through Britain at the expense of the indigenous red, *S. vulgaris*, is a familiar example. However, the situation is rather more complex than at first appears, and demonstrates the difficulties involved in the documentation of competitive exclusion. The red squirrel was, apparently, already subject to large fluctuations in numbers prior to the introduction of the grey (possibly as a result of a disease), and was at a low density before the main spread of the grey [82]. The red has since failed to recover in areas where the grey took over, and its range has continued to decline with extension of the grey's range since the 1930s. Their diets do overlap widely [51], but there appears to be some habitat differences between the two species, as the red squirrel persists longer in coniferous plantations and the grey shows some preference for deciduous woodland. Direct combat, the killing of young red squirrels by greys and the avoidance of grey adults by red squirrels have all been recorded [82]. This indicates the possibility of interference competition between the red and the larger grey species, but there is still debate as to whether the continued changeover of the two species is due directly to such competition.

30   *Community Structure and the Niche*

Control   All removed   *Larrea*   *Ambrosia*
removed   removed

Fig. 3.4 Diagrammatic representation of treatments used in manipulative experiments investigating the interactions between two co-dominant desert shrubs *Larrea tridentata* and *Ambrosia dumosa*. The above involve *Larrea* as test plant and similar treatments are used with *Ambrosia*. (After Fonteyn and Mahall [76].)

(b) Manipulation experiments have been conducted on artificial and natural plant assemblages. The four treatments used to investigate the interactions between two co-dominant desert shrubs *Larrea tridentata* and *Ambrosia dumosa* provide a good example of the manipulation technique (Fig. 3.4). Interference, in the form of allelopathic interactions (direct inhibition using toxic chemicals) and competition for water was clearly indicated among the plants in the area. Removal experiments on mown grassland fields have demonstrated that grasses are limited by one or more dicotyledon species and winter herbs by at least one perennial species [77]. However, selective removal of competitors of two sorrel species, *Rumex acetosa* and *R. acetosella*, from grassland communities produces a more complex picture of the niche relations between these and other members of the grassland community (Fig. 3.5). Similar examples of competitive exclusion have been found with other *Rumex* species [79].

Several manipulative experiments have demonstrated competitive exclusion in small mammal assemblages [80], *Plethodon* salamander guilds [81], and ant and rodent assemblages [64]. Manipulation of populations of flatworms, starfish, barnacles, limpets, and sparrows provide similar data [51, 57, 83, 296].

(c) The occurrence of interspecific competition has been inferred from negative correlations between the spatial distributions of species. Such relationships can lead to the so called 'checkerboard' distributions, where, for example, two competing species occupy islands or geographical areas to the mutual exclusion of each other in an irregular geographical array (Fig. 3.6). This can also occur between different taxon-

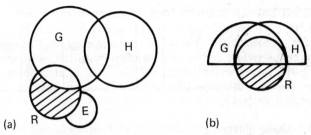

Fig. 3.5   Niche relationships between (a) *Rumex acetosa* and (b) *R. acetosella* and other members of the grassland assemblage. In each diagram the fundamental niches of grasses (G) and all herbs except *Rumex* species (H) are indicated as overlapping areas. Fundamental niches of *Rumex* (R) are continuous outlines, with the realised niches hatched. E is that portion of the fundamental niche of *R. acetosa* only expressed when grasses were eliminated. (After Putwain and Harper [78].)

omic groups, as shown by the reciprocal changes in biomasses of *Anolis* lizards and insectivorous birds which compete for prey [84].

Smaller scale spatial separation of species, which is not

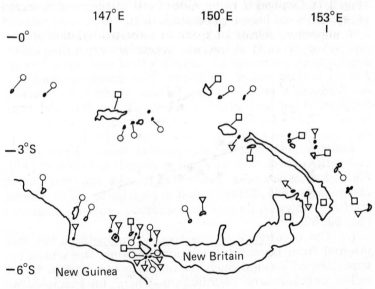

Fig. 3.6   Checkerboard distribution of *Pachycephala* flycatcher species in the Bismark region. □ - *P. pectoralis*, ▽ - *P. melanura dahli*, ○ - neither (After Diamond [213].)

based on habitat features, has often been described among small rodent guilds. Competition for space through inter-specific territoriality leads to non-overlapping ranges between two species, or reduced overlap in multispecies systems [80, 85]. Interspecies territorial defence is also thought to cause regular interspecific colony spacing among desert harvester ants [64].

3.4.2 Niche shifts

The argument used to imply that competition controls the above spatial distributions is that the species concerned may occupy mutually exclusive habitats in areas of sympatry (where distribution ranges of the species overlap) but occupy a wider range of habitats in the absence of other species (allopatry). For example, the epiphytic intertidal bryozoan, *Alcyonidium hirsutum*, occupies its greatest range of heights on plants of the seaweed *Fucus serratus* in the absence of its two potential space competitiors (two other bryozoan species). This range decreases when first one and then both competitors are present [46]. Similarly, mean niche width is an inverse function of the number of species in many plant assemblages (Fig. 3.7). Changes in niche width in the presence of potential competitors has been termed niche shift.

Competitive release of species in isolated habitats demonstrates niche shift in reverse, where an expansion of the

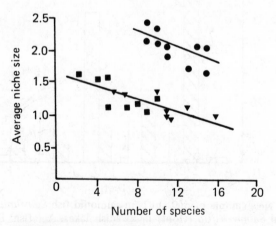

Fig. 3.7 The relationship between average niche size and number of species in samples from desert shrubs (■), grasslands (▼) and mixed forest (●). (After McNaughton and Wolf [6].)

realized niche towards the fundamental niche takes place in the absence of mainland competitors. The ecological literature is rich in examples of niche shift in other types of habitats, mainly involving animals. Figure 3.8 shows a clear example. Other interesting examples include dietary niche shifts among congeneric bumble bees [87], altitudinal [88] and microhabitat [89] shifts in salamander guilds,

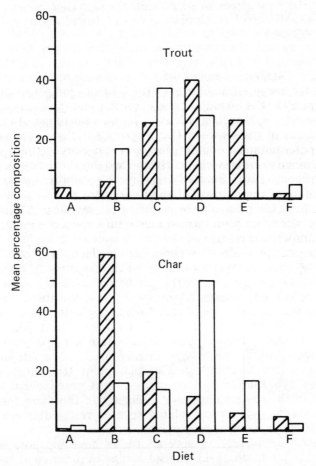

Fig. 3.8   An example of diet shifts in salmonid fish *Salvelinus alpinus* (char) and *Salmo trutta* (trout) in Swedish lakes. A — fish; B — small crustaceans; C — large crustaceans and molluscs; D — insect larvae; E — terrestrial insects; F — miscellaneous. Hatched bars — sympatric diet, open bars — allopatric diets. (Data adapted from Nilsson [50].)

habitat shifts in *Anolis* lizards [90], spatial and temporal shifts between prairie voles and cotton rats [59], and temporal shifts across taxonomic boundaries between humming birds and moths [91]. Many of these examples are just two species systems, but as discussed earlier, diffuse competition between several species is an important factor. Evidence of niche shifts under such conditions can still be found (e.g. [45]).

Further examples of niche shift through field observations and experimental manipulation can be found elsewhere [45, 51, 85, 296].

3.4.3 Character displacement
This is a mechanism related to the ecological displacement of niches described above. It can be defined as a process by which a morphological, behavioural, or physiological character state of a species changes under natural selection arising from the presence of one or more ecologically similar species in the same environment [92]. Character displacement is described between species which are very similar in the allopatric state, but differ in some character in sympatry. For example, changes in size or mouthparts can lead to differences in the size of prey utilized, which in turn, reduces competition and allows coexistence.

In order to identify character displacement, one must assume that the species originated in allopatry. In addition, a comparison with what the predicted character state of a species in the sympatric region would be in the absence of the competitor is necessary to validate that character displacement has actually occured. In a critical review, Grant [92] concluded that evidence for behavioural and physiological character displacement was satisfactory, but much of the evidence for morphological displacement (the most widespread type) was weak or simply represented clinal variations in the species character (a change in the characteristics of a species population related to a corresponding environmental change over a geographic area).

Since Grant's review several studies have appeared which provide much stronger evidence for the occurrence of character displacement. These include the displacement of snout-vent length in two species of subterranean skink (*Typhlosaurus* spp.) which is related to a dietary shift in prey size [93] and displacement in body size of two, small marine deposit-feeding snails (Fig. 3.9). Although the original

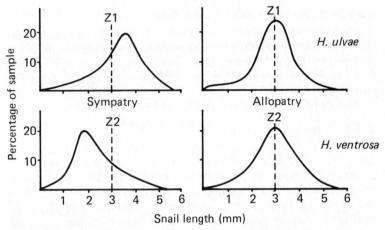

Fig. 3.9 Character displacement in the shell length of the marine deposit-feeding snails *Hydrobia ulvae* and *H. ventrosa*. (After Fenchel [94].)

prediction, that the size displacement leads to differences in particle size selection by the snails, has been disputed [95], the divergence in size is difficult to explain by any mechanism other than some form of interspecific interaction. Morphological character displacement has also been demonstrated in other invertebrates (e.g. [96, 97, 299]), but from a theoretical basis, character displacement based on food size is unlikely to be confirmed among small arthropods and a competitive gradient is predicted instead [98]. It is argued that larger predators are able to utilize food sizes unavailable to smaller predators, while the reverse is much less true. Character displacement based on food size may thus be confined to a narrow range of the larger invertebrates.

In order that character displacement can occur, there must be some initial difference in the resource spectra utilized by the species, otherwise they will compete severely and selection will eliminate one from the overlap zone. This can result in the continuous allopatry seen in the examples of 'checkerboard' distributions. At the other extreme, the species may be so different at the time of contact that extensive sympatry will be established without noticeable character displacement [92]. Theoretical models indicate that significant character displacement may only occur when phenotypic variances of the species are constrained to sufficiently small values [99].

This provides some relationship between the Niche Variation hypothesis and studies on character displacement. Lastly, character displacement is only likely when a single environmental resource dominates the ecology of both species. If two or more resources are important, differential resource utilization can obviate the necessity of morphological character displacement in sympatry (see Chapter 4).

3.4.4   Changes in resource levels

Traditionally, studies demonstrating the impact of competition on community organization have been based on comparisons between similar habitats possessing differing degrees of species packing. Changes in species activities are then described as responses to changes in the intensity of competition. A more subtle method is to document the outcome of direct changes in resource levels within a single guild or assemblage. Competition is shown to be important if a decrease in species number follows a reduction in resources, and, more importantly, if this reduction has the greatest effect on the species apparently suffering the most intense competition. A number of studies have documented such changes in species composition. For example, as resource levels (number of microhabitats) decrease on a north—south geographical gradient, foliage-gleaning birds show a corresponding decline in species number [38]. In addition, the species suffering the most intense competition was the first to drop out, and the species experiencing the least interspecific competition showed the most variable morphological structures associated with resource use (as predicted in the Niche Variation hypothesis). The impressive studies on Darwin's ground finches in the Galapagos Islands [100] and tropical stream fishes [101] both demonstrate reductions in species numbers with seasonal reduction in food availability. In the same way, when food resources change from year to year, granivorous desert rodents are excluded from areas in regular order [102] and some auk (Alcid) species are 'squeezed out' of their normal foraging niche [103].

Overemphasis on occasional food shortage obscures another important aspect of community organization involving the immigration of competitively inferior, fugitive species during intervening seasons of resource abundance [63]. In many bird and mammal guilds, some species never face prolonged food shortage as they migrate between flushes of food in different habitats. Many nectar-feeding birds from Trinidad and Tobago [63], Darwin's ground finches [100] and other island bird species [85] provide good examples.

3.4.5 The Niche Overlap hypothesis

As discussed earlier, this hypothesis predicts increased separation of niches within a community either as species diversity increases or resource levels decrease. This is presumably accompanied by a related decrease in the size of the realized niche of constituent species as a result of competitive interactions.

There is widespread evidence demonstrating a decrease in niche width as food resources decrease. For example, many species of tropical stream fish change from small, distinct food niches during the relatively poor, dry season, to broad and widely overlapping niches in the more productive wet season [101]. Similarly, in the Galapagos ground finches, the mean number of plants selected per species fell from 11.8 to 4.4 during the poor dry season [100]. The reverse pattern of increasing niche width and overlap with increasing prey availability has been documented in predatory Erpobdellid leeches [105].

The decrease in niche width with increasing species diversity has been mentioned in Section 3.4.2, and many specific studies have shown the mean spatial niche width of plants to be a negative function of species richness [106].

The predicted decrease in niche overlap with increasing species richness (leading to an increase in niche separation) has been shown for lizard guilds on three continents [41, 107], ground-dwelling small mammals [106], desert rodents [108], foliage-gleaning birds [68], ants [109] and tropical stream fish [101]. These studies are all based on a pairwise comparison of niche overlap.

The generality of the Niche Overlap hypothesis has been questioned [86]. Some small mammal assemblages, for example, show a positive relationship between niche overlap and species diversity whilst the niche width itself remains unchanged [61]. Such assemblages are probably unsaturated, and the shared resources, such as space, in the above example, are not limiting [106]. The inclusion of additional species in these situations may take place through an increase in overlap, as competition for the resource is not intense.

Theoretical studies also disagree on the nature of the relationship between niche overlap and diversity. One ap-

proach, using Lotka—Volterra based models, predicts that
maximum tolerable niche overlap should increase with more
intense competition [110]. In contrast, simulation studies
testing the relationship between community stability and
diversity support the Niche Overlap hypothesis [7]. It is
found that as the number of system components increases,
the mean interaction intensity (comparable to niche overlap)
must decrease if the system is to remain stable. Data from
plant and animal assemblages support these predictions
(Chapter 10).

The decrease in niche overlap as diversity increases can be
explained by stronger diffuse competition. In foliage-gleaning
bird guilds [38] and lizard assemblages [41], total (diffuse)
niche overlap increases significantly with species diversity,
even though average overlap between pairs decreases. This
adjustment of overlap with species diversity could potentially
result in a relatively constant level of interspecific inhibition.
High overlap with few competitors could be equal to lower
overlap with more competitors, so that the actual intensity
of interspecific competition per species could be similar in
communities of widely divergent species densities [41].

3.4.6 Competition in the past

It is possible that past competition may have been important
in organizing a present day community, even though direct
experimental manipulation does not reveal the expected
changes in niche size and patterns. A species that continually
experiences strong interspecific competitive pressure, is
presumably unlikely to retain the capacity to exploit many
resources over ecological time, whereas a population periodically released from interspecific competition naturally, might
[3]. For example, the distribution of two small desert rodents
has been related to past competition, which caused strong
selection against the utilization of similar habitat types [44].

The taxonomic relationship of species within the community offers another potential way of exploring the importance of past competition. As congeneric species should be
strong competitors, one might expect fewer pairs of congeners
in natural communities than in a sample from various species
and genera over a broad geographical area. Instead, several
studies have found that as species number increased, there
was either no impoverishment of congeneric pairs or more
than expected [9, 311]. However, some indirect support has
been found in lizard assemblages which show similar generic

diversity in three geographic regions [101] and in some studies on lake zooplankton, which showed that congeric species niches were the furthest apart [74].

3.5 Conclusion

Both intra- and interspecific competition have been shown to influence the species niche. Intraspecific competition causes niche expansion, and pushes the realized niche towards the fundamental one (or beyond). Interspecific competition tends to counteract this expansion and also plays an important role in species relations. Where species are very similar, competitive exclusions can occur, but by reducing overlap and the size and/or position of their niches through niche shift or character displacement, such exclusion can be avoided and coexistence results.

The two types of competition have opposite effects on niche width, but can we discern which has the greatest influence in the organization of communities? The examination of nearest neighbour distances between individuals of the same and different species provides one method of comparison. Results from field experiments on desert shrubs indicate that between-species interference is usually more intense than that within-species [54, 76], although the reverse has also been reported [54]. Similar studies on limpets have shown either greater effects from other species or no difference at all between intra- or interspecific pairs [55]. Monitoring the effects of a reduction in food availability offers another method of approach. If interspecific competition determines the partitioning of resources, then species should diverge in diet, as found in ground finches [100] and squirrels [104]. If intraspecific competition is more important, optimal foraging theory predicts that an increase in diet width should occur, as seen in several humming-bird species [48]. However, observations of ecological release and the fact that most species do not occupy their fundamental niche, both suggest that interspecific competition has a greater influence on niche size. It has been suggested that at equilibrium, intraspecific competition is balanced by total interspecific competition [9], but the balance depends on both the resource levels and degree of diffuse competition so that under different conditions, either one or the other may dominate.

Chapter 4

Competition
and the niche;
limiting similarity
and differential
niche overlap

In the previous chapter, competition was shown to influence both the size of the niche and the degree of niche overlap between species. The ideas discussed there lead to an interesting question. If no two species can occupy the same niche indefinitely when resources are limiting, then how similar can two species be and still coexist? Restating this in terms of niche theory, how much overlap can occur between niches without competitive exclusion? This is the problem of limiting similarity.

4.1 Limiting similarity — the theoretical approach
The concept was first examined by MacArthur [111] but the theoretical approach has since been developed more rigorously to specify quantitatively just how similar coexisting species could be. This type of work has, for the sake of simplicity, been based on Lotka—Volterra type competition models [112] and species separation along an unidimensional resource spectrum (Fig. 4.1). Each species in the system uses

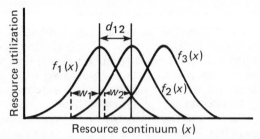

Fig. 4.1 Theoretical resource utilization by three species with similar utilization curves (fx). d = distance apart of the means; w = standard deviation of utilization.

the resource in a way described by their resource utilization function (fx), and the degree of niche separation is characterized by the niche separation ratio $\rho_{ij} = d_{ij}/w_{ij}$, as described in Section 2.3.2. This ratio is large for well-separated niches, and small for highly overlapping ones.

Early models assumed a particular and constant shape of utilization function (e.g. bell-shaped), whereas some later models allowed random fluctuating components in the competition equations (i.e. fluctuations in the carrying capacity of a population) [7]. In all cases, a ratio $d/w > 1$ is a prerequisite for adjacent species to persist together without putting too many constraints on the model's assumptions or flexibility. Another approach has been to allow the shape of the utilization functions to 'evolve', such that the system achieves maximum fitness of constituent species [113]. These models predict that the system tends to reach a situation where $d/w > 1$ for adjacent species on the resource spectrum.

An alternative method considers theoretically the survival chances of a species attempting to wedge its niche between two established species [114]. For bell-shaped resource utilization functions, MacArthur and Levins [114] found that the limiting d/w ratio for successful invasion is 1.56.

On a theoretical basis, there is a limit to niche overlap, and coexistence on one resource gradient is only likely when the average difference between species exceeds the typical variability within species i.e. $d > w$. Typically, a ratio $d/w \leqslant 1$ indicates potentially strong competition for the resource, while a ratio $d/w > 3$ infers that there is no interaction between species, i.e. no overlap [29]. Can we see such a degree of limiting similarity in nature?

4.2   The degree of limiting similarity

Due to the difficulties in obtaining a reliable estimate of actual resource utilization and of comparing communities in different habitats, it has often been convenient or necessary to estimate resource differences by using a species characteristic that indicates its position on a resource dimension. As such, morphological characters have been most widely used, because of the ease and objectivity of obtaining measurements and the independence of such measurements from habitat structure [107]. This involves measuring the average distance between nearest neighbours in morphological space (e.g. species with the closest body length), in order to obtain some measure of the degree of species packing and hence, limiting similarity. The method assumes that morphological distance bears a consistent relationship to ecological similarity. Most studies have used the size of feeding or trophic structures, body size or weight of the consumer as a morphological indicator of food size utilization. This is acceptable providing that the variation in size of the consumer is correlated with variation in size of its food, i.e. the larger the character, the larger the food resource utilized. Such correlations have been demonstrated many times for vertebrate consumers (e.g. lizards [115, 116], birds [57, 117], and mammals [102, 104, 118]). but infrequently for invertebrates [119, 120].

The following discussion is mainly based on the food size resource dimension, and is thus confined to animals, but the conclusions can equally apply to use of any other resource in both animals and plants.

4.2.1 · Hutchinson's rule and Dyar's law

In his classic paper [121], Hutchinson demonstrated an apparent constancy in the magnitude of morphological character divergence in the ratio of mouthpart sizes among a few co-existing, congeneric species of insects, birds and mammals. Values ranged from 1.1 to 1.4, with a mean of 1.28, which was tentatively suggested as the lower limit to similarity for competing species. Since then, many workers have come to recognize a 1.3 size ratio, or a corresponding weight ratio of 2, as a biological constant or 'rule' among groups of sympatric species, especially in vertebrate and invertebrate feeding structures or body size. This list of supporting evidence is impressive and includes spiders [120], tiger beetles [119], lizards [115], salamanders [89], squirrels [104], bats [118] and desert rodents [102]. Bird studies provide some striking examples. For example, bill length ratios between sympatric congeneric species from 46 temperate and tropical families cluster around 1.3 (range 1—1.7, mean 1.15 [103]). Studies on eight species of *Ptilinopus* and *Ducula* fruit pigeons in New Guinea also demonstrate this 'rule' (Fig. 4.2). They are graded along a size sequence over a 16-fold range in weight, and the mean weight ratio between adjacent species is 1.9 (range 1.33—2.73). The species separate through size of food and feeding position on different parts of the branches which could support their weight.

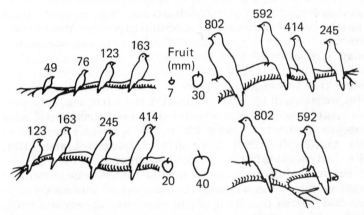

Fig. 4.2   Size separation in fruit-eating, coexisting, New Guinea pigeons (*Ptilinopus and Ducula* spp.). Bird weights (g) are given above each species. (After Diamond [57].)

A similar relationship has commonly been described between the linear dimensions of successive insect instars, referred to as Dyar's Law, where one instar is approximately 1.3 times larger than the preceding one [115, 122, 124]. Similarly, successive age classes of salamanders fit this law [89, 125]. It is argued that such examples represent a developmental response to potential competition between successive age classes or instars within a species.

Some authors feel this empirical relationship is not understood [122] but Maiorana [125] has offered a possible explanation for both its constancy and magnitude. If ecological segregation of two species (or age classes) requires a minimum level of overlap in their frequency distributions, and if the species show a similar degree of morphological variability, then the linear displacement in mean size will also be relatively constant. For many structures, the coefficient of variation (CV = standard deviation as a percentage of the mean) tends to range between 4 and 10, and most frequently between 5 and 6. A CV of 5.5 for two species with a size ratio of 1.28, will produce an overlap in their frequency distributions of only 1—2% [125]. This explanation also predicts that Hutchinson's 'constant' should be somewhat variable (as found in the literature), for the morphological variability of any species is not absolutely constant.

Thus, in a wide range of situations, species seem to be organized in a non-random fashion along a resource, and it appears, on this basis, that competition has produced a regular distribution of body or feeding structure size among many sympatric species.

4.2.2 *Criticisms of these rules*

The widespread use of Hutchinson's 'rule' has been criticized on many occasions. It has been pointed out that the 1.3 ratio appears to describe series of recorders, tools and bicycles and may well derive from generalities about assembling sets of tools rather than being a biological peculiarity [122]. However, such similarity may have to do with our perceptual abilities in the construction of these tools, a cause applicable to, and perhaps derived from, the natural world [125]. Roth [126] criticizes data-collecting methods in certain studies, and points out that many size ratios change seasonally. Also, the ratios themselves might be size dependent, as allometric growth can cause ratios of measurements within the animal to change with an increase in size [126]. This doesn't explain the average magnitude, and simply moves the problem one step back.

More importantly, many studies have demonstrated a great overlap in sizes and kinds of foods used, in spite of large differences in body size or size of feeding structures. Adjacent species of some raptorial spiders, for example, show $d/w > 1$ for body size, but $d/w < 1$ for prey length [127]. Seed-eating rodents often show large overlap in seed sizes eaten, despite weight ratios of 2 or 3 [102]. Similarly, large predators generally utilize food unavailable to small predators, whereas the reverse is much less true. This indicates that consistent size differences along a range of prey sizes might not lead to a consistent reduction overlap between adjacent species [98]. Energetic considerations may also be important here.

In these examples, morphological variation is considerably less than variation in use of the resource, so complete separation in morphology may be necessary to allow sufficient separation on one resource dimension. This may also apply to a wide variety of other data [125] where the magnitude of morphological separation of two size displaced species or age classes is directly proportional to their morphological variability, and usually lies between 3 and 8 standard deviations of the means (of logarithmically transformed normal distributions). The concept of limiting similarity still applies, but in these situations a greater degree of separation than theoretically predicted is again necessary (3 −8 times greater for these data). The large variability in resource use can, in part, also be explained by the plasticity of the behaviour of the species. This may be associated with the fact that food items are often not selected just on size, but also on energetic returns, ease of capture, taste, nutritive status and food abundance (e.g. in fly catchers [128, 129]). These ideas are put forward in optimal foraging theory (e.g. [40]).

Finally, the generality of Hutchinson's 'rule' has been questioned when several species are found to coexist while utilizing similar types of food, yet do not show differences in body or feeding structure size of the predicted magnitude, if at all (e.g. [103, 118−120, 124, 130]). How can such findings be explained if there is a limiting degree of similarity as proposed by theoretical studies and suggested in Hutchinson's 'rule'?

4.2.3 Applicability of these rules

Obviously these rules describing the degree of limiting similarity do not apply over the whole range of organisms or situations. The theoretical models assume that species compete for resources distributed along a unidimensional gradient, so any fit to these models suggests that this is how the particular species are dividing up resources. The supporting data come mainly from studies on sympatric, congeneric species that devote considerable time and energy to searching for and pursuing similar kinds of particulate foods. If size of the morphological character (through some relation to food) is the only significant niche parameter, one must expect at least a 28% difference between coexisting species. If the assumptions in the theoretical models are not fulfilled (namely limiting resources, saturated communities and separation on only one resource dimension), then we will not be able to identify the same degree of limiting similarity, if any. When we find the expected size differences but widely overlapping resource use, or coexisting species showing no differences in the morphological character examined, then it implies that this character may refer to an ecological activity secondary to the separation of two species. The expected degree of limiting similarity in trophic or feeding structures may not apply to many plants, or to those animals that harvest relatively abundant food (e.g. grazers or filterers), as they carve up nutrients with little time and energy in locating and harvesting them. For example, in guilds of rodent granivores, frugivores and carnivores, interspecific competition produces clear patterns in body size displacement and food particle utilization. In contrast, rodent grazers of closely related species rarely coexist in the same habitat, as large body size differences are probably not sufficient to permit coexistence [58].

We are thus left with the concept of limiting similarity being demonstrated theoretically, but the equivalent degree of limiting similarity is only seen within an assemblage or guild which satisfies the assumptions of the theoretical models. Most evidence has been found when some aspect of food has been the major limiting resource and in such cases, Hutchinson's 'rule' is applicable. When variation in resource use is greater than variation in morphology, then a greater separation than predicted is necessary for coexistence. When more than one resource is of importance, or resources are not limiting and the community is not saturated, then the assumptions of the theoretical models are not met. Species then appear to be much closer in morphology or more importantly, utilization of one resource, than expected, and the predicted degree of limiting similarity on any one resource is difficult to demonstrate.

This leads us on to two other important concepts in the consideration of the role of competition in community organization, namely niche dimensionality and differential niche overlap.

4.3 Niche dimensionality and differential overlap

As we have seen, most of the early ideas and a majority of the theoretical approaches to competition and its effects on niche overlap, have considered species using essentially just one resource dimension. The previous section showed a substantial body of evidence demonstrating that competition can lead to a limiting similarity between species on such a dimension. If this is a general biological rule, then as the number of competing species increases within a community, they will have to separate on more and more dimensions to preserve a minimal total resource overlap and level of competitive inhibition (as found in some studies described in Section 3.5). Thus, in studying community structure, one should clearly examine overlap in terms of Hutchinson's n-dimensional hypervolume (see Chapter 2), looking at all environmental gradients and resources utilized by the constituent species. Although this is conceptually powerful, the idea is too abstract and too difficult to apply, as one can never satisfactorily identify all the factors impinging on a species.

This idea of niche dimensionality is important, however. As indicated earlier, coexistence between similar species can occur when their niches greatly overlap in one resource dimension but are separated substantially on another. This is known as differential niche overlap (Fig. 4.3). Depending on which dimension is being examined two species could be considered as showing high niche overlap and hence competing, or as being completely separated in niche space. Such problems have contributed to the difficulty in validating the competitive exclusion principle under natural conditions. Rather than the observed coexistence between two very similar species refuting the principle, it may simply mean that some significant ecological difference has been overlooked.

There is now a large amount of documented evidence illustrating how similar species coexist, and hence supposedly avoid competition, through differential niche overlap on two or more environmental gradients. Zoologists and botanists have approached the problem in rather different ways, reflecting, on the whole, the differences in the way species of animals and plants partition their resources. As such, it is simplest to discuss examples and general principles of each in turn. The differences in the spectrum of examples reflects a true bias in the ecological literature.

4.3.1 *Differential overlap in animals*
To demonstrate the concept, we can see many coexisting

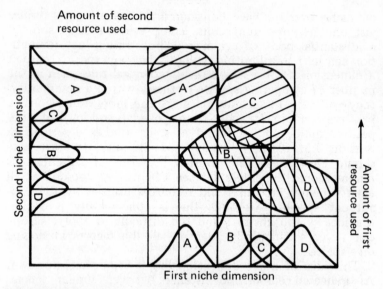

Fig. 4.3   Differential niche overlap amongst four species (A—D). Pairs with substantial overlap on one niche dimension can reduce competition by differential resource utilization on another. The hatched areas show the resultant niche space occupied by each species.

species showing a large overlap in food or space, but differences on a temporal basis. Such coexistence is possible providing the resources at one instance are relatively unaffected by what happened previously. The clearest examples are found in the coexistence between diurnal and nocturnal animals, such as hawks and owls, swallows and bats, grasshoppers and crickets. In coral reefs, many species of fish are active by day, different species are active around dusk, and fully night-active fish follow, all sharing similar foods and refuge spaces.

When coexisting species utilize similar food at the same time, they may be separated horizontally through the habitat, as seen in many species of woodland bird (Fig. 4.4). When coexisting species utilize similar food in the same areas at the same times, they may be separated along the vertical dimension of the habitat (Figs 4.4 and 4.5).

Complex resource partitioning is found in assemblages with many similar types of species. Two examples are given in Fig. 4.6.

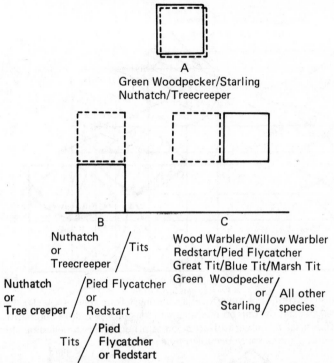

Fig. 4.4 Mechanisms which separate woodland bird species. A — food specialization without spatial separation; B — vertical separation; C — horizontal separation. (After Edington and Edington [131].)

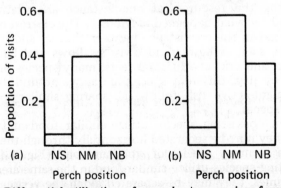

Fig. 4.5 Differential utilization of space by two species of waterboatmen which have large dietry overlap. (a) *Notonecta obliqua*, (b) *N. glauca*. NS — perch sites near water surface; NM — perch sites in the middle of the water body; NB — perch sites near the bottom of the water body. (After Giller and McNeill [132].)

(a)

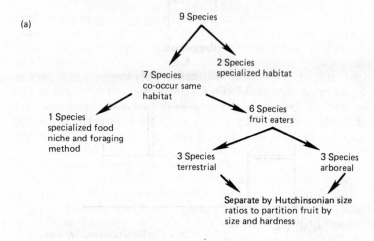

(b)

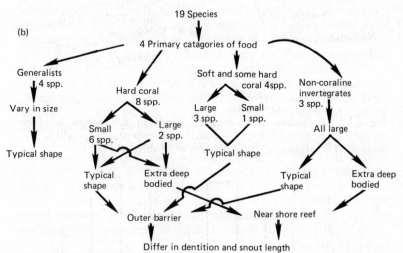

Fig. 4.6   Resource partitioning within (a) African rain forest squirrels [104]; (b) a coral reef fish assemblage [134].

Table 4.1 presents selected references to similar studies, covering a wide range of animal groups, and indicating the main resource dimensions on which the coexisting species are separating. Most examples involve resource partitioning within groups of taxonomically related species utilizing similar resources but a few have investigated the use of similar resources by widely differing taxonomic groups.

There is no doubt that differential overlap plays a very important part in the structure of communities, but can we identify any particular resource dimensions as being consistently more important than others? According to several authors, we can. Whittaker [3] suggests the niches of a passerine bird guild can be characterized by three major axes; vertical height of feeding and nest site, food size and food composition. Cody [135] suggests that vertical height, horizontal habitat and food type are the most important dimensions for grassland birds, and Brown [58] estimates that coexistence between species of granivorous desert rodents is achieved mainly on two niche dimensions; seed size and foraging area. An extensive analysis of 81 investigations of resource partitioning has been carried out by Schoener [136]. With no limit on the number of dimensions investigated, the mode for the number separating species was three, and where a maximum of three dimensions were investigated, separation was usually on two. Five major categories of species separation along complementary resource dimensions were also identified. (a) *Food type and habitat* — species showing overlap in habitat eat different foods, or vice versa; (b) *food type and time* — species use the same food at different times or different foods at the same time; (c) *habitat and time* — similar to (b); (d) *habitat and habitat* — species overlapping in horizontal habitat often differ in vertical habitat or vice versa; (e) *food type and food type* — species may eat similar-sized foods but different species or types, or vice versa. Examples of each of these categories can be found in Table 4.1. In addition, a number of major generalizations can be made on the relative importance of certain resource dimensions in partitioning species from studies which suggest that competition is the major organizing factor [136].

(a) Habitat dimensions are important more often than food type dimensions, which are important more often than temporal ones.
(b) Temporal partitioning (seasonal and diel differences between species) is rapidly exhausted as species number in the guild or community increases. (There may, in fact, be a greater degree of limiting similarity on a temporal basis than on the food dimensions discussed earlier).
(c) Given this, predators separate more often by being active at different times of the day than do other groups. This

Table 4.1 Differential resource overlap within selected animal guilds and assemblages

Animal group	Overlap	Separation	Reference
Invertebrates			
Terrestrial			
Tiger beetles	H, F	ST, HD	[119]
Raptorial spiders	F, S, FB	T, H	[127]
Orbweb spiders	F	T, S, HV	[130]
Soil collembola	F, T, H	HV	[154]
Periodical cicadas	F, T	FH, HD	[155]
Grasshoppers	F, H	HM, P, T	[156]
Leaf hoppers	FH	HV, HD	[157]
Tropical cockroaches	H	T	[158]
Rainforest ants	T, H	HD, HF, F	[159]
Aquatic			
Freshwater zooplankton	H	T, HV, F	[74]
British notonectids	F, H	FB, P, PL	[132, 161]
Hydropsychid caddis	FB, H	HD, P, HM	[160]
Calanoid copepods	HD	HV, T, S	[162]

Vertebrates			
Stream fish	HF	HM, ST, FB	[101]
Streambank salamanders	H	S, HM, FS	[89]
Desert lizards	H, S	F, HD	[41]
Desert lizards	F, H	T, HM, FH	[107, 151]
Anolis lizards	HD	S, HV, FS, HM	[136]
Cormorants	F, H	S, HF, HB, T	[103]
Auks	F, H	FB, HF	[103]
Flycatchers	FB, H	HV, HD, HF	[133]
Forest gleaning birds	HF	FB, FS, F	[145]
Bats	F	FS	[188]
Desert rodents	F, H	FS, HD, HF, FB, PL	[58, 102, 151]
Desert rodents	FS	HF	[151]
Across taxonomic boundaries			
Desert ants and rodents	H, F (Seeds)	FS, HD	[64]
Hummingbirds and insects	F (Nectar)	FH, H, T	[91]
Insects, birds and mammals	H, F (Conifer Seeds)	HD, HV, FB, T	[163]
Freshwater caddis and alderfly larvae	H, F (Prey)	FB, T	[164]

KEY: F, Food type; FH, Host/Prey species; FS, Food size; FB, Foraging behaviour; H, Habitat type; HM, Microhabitat; HD, Horizontal distribution; HV, Vertical distribution; HF, Foraging area; S, Size; ST, Size feeding apparatus; T, Time/Season; P, Physiology; PL, Locomotory efficiency is mostly due to their prey showing their own daily activity patterns (in effect creating independent resources). This possibility does not apply in general to herbivores for instance

(d) Terrestrial poikilotherms frequently partition food by being active at different times of the day.

(e) Animals with long generation times cannot partition the year as finely as those that mature quickly. Vertebrates, in general, segregate less by seasonal activity than do lower animals.

(f) Segregation by food type is more important for animals feeding on large food in relation to their own size than it is for animals feeding on relatively small food items. This has also been suggested on a theoretical basis [98].

4.3.2   Niche dimensionality in plants

Specialization along niche axes does not seem to have evolved to the same extent in plants. In general, all green plants have essentially similar requirements for food, water, carbon dioxide and a basic set of mineral nutrients. On the grossest global level, the 300 000 terrestrial plant species may have just 30 different limiting resources [295]. Thus, there is little opportunity for diversification along the food axis [137, 138], in sharp contrast to animals. Plant species populations tend to overlap broadly along environmental gradients [11], and sharp plant species distribution boundaries, explained on the basis of epidemic models [139], tend to be the exception. Despite such overlap, each plant species probably has its own population centre, differing from those of other species. Habitat space is divided on the basis of adaptations along gradients of quantitative variation in environmental factors (such as light, water and nutrients), as well as being associated with variations in elevation, slope, aspect, soil type etc. In the broad sense, these will explain plant distributions and some measure of reduced resource overlap and coexistence of species in an area. However, one still finds many species of plants together in small areas of habitat where differences in physical factors may not seem great enough to explain such coexistence. As green plants are undoubted competitors of light, water and nutrients [140, 300] and probably space, subtle differences in requirements allowing coexistence in the face of apparently large resource overlap should be evident.

One such mechanism has been suggested by Tilman [295].

He demonstrates that it is theoretically possible for numerous plant species to coexist at equilibrium in a spatially hetero-genous environment if the species differ in the proportions of various nutrients that they require and are competitively superior over only a small range of resource supply ratios. A substantial body of data is presented in support of this idea [295].

Most plants are fixed and rooted, and a variety of vital activities, which in animals are accomplished by patterns of behaviour such as search and choice of food, mating, competitive interactions with neighbours, etc., are achieved in higher plants mainly by variations in form and life history [2]. Thus, plant ecologists tend to investigate how species share resources (and hence community organization) by investigating the physiognomy of the community (its physical structure) and different strategies of life history among the plants. From these, one may infer niche differentiation in much the same way as animal ecologists have used morphology. Differences in growth form among plants are therefore thought of as visible indications of niche differentiation.

A complex example of such coexistence of plant species can be seen in the Sonoran Desert of south east Arizona [3]. The plants show differentiation in the heights at which species bear the greater part of their foilage (Table 4.2) and in the seasonal behaviour of foliage. This ranges from per-sistent evergreen species, through semi-deciduous species, to species with short-lived leaves and finally cacti which lack leaves altogether. This gradient of decreasing leaf persistence and increasing stem and branch photosynthesis must contri-bute to reduced competition. Other environmental gradients, such as degree of shading tolerance, differences in root

Table 4.2 Differences in the height at which Sonoran desert plants bear the greater part of their foilage (adapted from Whittaker [3])

Height above ground of greatest part of foliage	Plant type
Near zero	Herbs with stems and leaves trailing
Few centimetres	Other herbs
Few decimetres	Semi-shrubs
0.5–2 m	True shrubs
2–5 m	Arborescent shrubs
6–9 m	Giant cactus

depth and form, are also important. Together such differential resource overlap and utilization leads to the striking
diversity of plant species in this desert.

As indicated in this example, niche differences in plants
also involve a wide range of functional relationships in physiology, life cycle and adaptation to other species, only a few
of which will have any visible expression in growth form.
However, to overcome such problems in the analysis of niche
differences, there have been attempts to identify primary
life history strategies, distinguishable with respect to their
response to shortages of light, water or nutrients (as may
occur in the presence of competing species or poor conditions). Grime [141] describes three broad primary plant
strategies. *Ruderal species* ensure production of seeds at the
expense of vegetative growth, and tend to die once seeds are
produced. This is an important strategy for partitioning
resources between parent and offspring, especially when
resources, such as space, are limited. *Competitor species*
maximize capture of resources, through rapid turnover of
leaves and roots and continuous spatial rearrangement of
absorbtive surfaces, allowing the plant to adjust to changes
in distribution of resources during the growing season.
Finally, *stress tollerators* are adapted for endurance, allowing conservation of captured resources so as to exploit
temporal variation in availability of resources in unproductive habitats. A simple example is found in the coexistence
of two annual grasses which dominate the Great Central
Valley of California [142]. *Erodium botrys* has rapid root
penetration as a seedling (a stress-tolerator strategy), so is
favoured in competition for soil nutrients. *Bromus mollis*
grows taller and becomes the superior competitor for light,
shading out *Erodium* if adequate nutrients are available (the
competitor strategy). Coexistence over time thus occurs, as
Erodium is favoured in poor soils and drought years, whereas
Bromus is favoured in better soils and normal years.

Regenerative strategies are also important in avoiding
resource overlap, reducing competition and allowing the
exploitation of unused space provided by the death of
adults or by disturbance in the community [78, 137, 138].
Five major types of regenerative strategy have been recognized [141].

Thus the use of many different growth and regenerative
strategies can allow coexistence of plant species in an area
over time. Further details can be found in the authoritative

books of Harper [75], Miles [4], and other relevant works [143, 144].

Plants seem to partition the temporal niche dimension to a much greater extent than animals. This is possible as the use of many plant resources at any one time does not affect resource availability at another (e.g. light). Both temperate and tropical forest plants have evolved towards diverse ways of flowering and fruiting, through differential adaptations to the seasonal cycle and to animals and wind for pollination and dispersal. For example, in temperate deciduous forests, spring beauties (*Crocus vernus*), hepaticas (*Hepatica* spp.), dog's-tooth violet (*Erythronium dens-canis*) and other herbs develop foliage and flower early before the trees are in leaf. Other groups have maximum growth and flowering in spring and summer, whilst others, like asters (*Aster* spp.) and golden rod (*Solidago* spp.), grow and flower later in summer and autumn [3].

Lastly, an important mechanism, through which wide overlap occurs but competitive exclusion is avoided, was documented by de Wit (see [138]). In mixtures of two species of grass, or a grass and a legume, intraspecific competition was greater than interspecific competition. This sets up a frequency-dependent situation, which maintains a self-balancing equilibrium state of coexistence (case 4 of the well-known competitive models based on Lotka—Volterra equations [52]).

4.3.3 *Differential overlap within a species*

Competition for limiting resources can occur within a species, and as individuals will be using essentially the same niches, such competition is likely to be most intense. Avoidance of competition through polymorphisms or continuous ecological variation within a species may occur (see Section 2.2). Under certain circumstances, conspecific individuals occupy different subniches or adaptive subzones, subdividing and perhaps expanding the total niche or zone utilized by a species population [56]. Such conspecific resource partitioning is frequently demonstrated in sexual dimorphism, leading to differential niche utilization by the sexes.

In birds, morphological and ecological sexual dimorphism is frequently expressed in the size and structure of feeding apparatus and differential foraging behaviour. Among certain species of the West Indian woodpecker genus *Centrus*, differences in the size and shape of tongue and beak are linked

to differences in foraging localities on trees and prey s̄̄̄
taken by the sexes [56]. Spatial differences are also comm̄̄̄
as in Henslow sparrows [*Passer herbulus henslowii*] wh̄̄̄
males use peripheral parts of the pair's feeding territō̱̄
whilst females use interior portions [145]. Similarly, vertic̄̄
differences in foraging zones between males and femalȩ̄
have been demonstrated for several foliage-gleaning species̄̄
such as vireos [146] and spruce warblers [147]. Differencē̱
in the microhabitat distribution of sexes has also beeñ̄
reported in desert rodents [148].

Niche partitioning between sexes of dioecious plants hā̱
also been found. In spinach (*Spinacia oleracea*), males hav̄̄
rapid growth and early flowering, but tend to senesce and dī̧
quickly. Females continue to grow and develop, and usȩ̄
resources freed by the death of the males to set seeds and
increase their reproductive vigour [149]. The opposite
temporal niche partitioning is found in the campion *Silene
dioica* [150] with females predominant in June and males
rising to equal abundance in August. Spatial niche partitioning
has also been reported for dioecious temperate and tropical
plants [150]. Dog's mercury, *Mercurialis perennis*, shows
spatial partitioning on the basis of soil pH preferences, and in
Trophis involucrata, a small tropical understory tree, the
sexes separate on the basis of soil phosphate levels.

Such ecological differentiation in animals and plants, may
have had a role in the evolution and maintenance of sexual
dimorphism and dioecy. The actual benefit to the species
could have been in response to two different selection
pressures; (a) reduction of intersexual competition, or (b)
increase in interspecific competitive ability. In this latter case,
as reproductive labour is often divided between the sexes and
resources are also divided, the habitat is more fully occupied
(i.e. a larger niche is occupied). This fits in with the Niche
Variation hypothesis discussed in Chapter 2.

4.4 Conclusion

On the basis of an immense body of literature, some of which
has been examined in these last two chapters, many ecologists
have concluded that competition for limited resources has
been the primary determinant of species packing and hence
species diversity [136, 301]. Intense exploitation competition
can lead to one species out-competing another for limiting
resources and interference competition can prevent access
to sufficient resources, leading to exclusion of one or more

mpetitors. Each species will have its own, slightly different,
ource preferences and uptake capabilities. As resources
:ome scarce, only the most efficient users will prevail,
suring that resource utilization by the community will
: maximal at equilibrium [28]. Over a short, ecological time
:ale, this could result in a decrease in species eveness and
ventually species number. Over longer periods of time, in-
:reasing interspecific competition is countered by selection
'or specialization and differential niche overlap. Coexistence
)f increasing numbers of species can then be maintained as
 result of competition maintained niche diversification up
 o some theoretical limit, as a function of either the number
)f discrete resources present, or maximal tolerable niche
overlap or both.

Niche separation is generally multidimensional, and as the
number of competing species increases, they will have to
separate on more and more dimensions in order to preserve
the minimal resource overlap necessary to alleviate competitive interactions in a saturated community. Thus, similarity in one dimension is usually, but not always, correlated
with distinctiveness along a complementary one. It appears
that two or three major complementary dimensions can be
identified, along which species in a guild can separate in niche
space.

One can obtain a measure of multidimensional overlap in
a number of ways [29, 41, 114, 151]. MacArthur and Levins
predict that overlap between adjacent species of less than
54% is necessary for coexistence without competition [114],
and similar values are reported elsewhere [152]. In most
studies where competition is considered to be the major
structuring mechanism in the guild, most species pairs show
overall overlap below this figure (e.g. [66, 89, 151, 153]).
Through the use of principal component analysis, one can
come nearer to a representation of comparative niches of
species in terms of Hutchinson's *n*-dimensional space [74,
154]. In such studies, clearer separation of species is apparent,
and this further demonstrates that complementarity of niche
dimensions tends to preserve minimal resource overlap and
reduces competitive interactions.

However, many authors believe that competition is not
the only biotic factor which can structure a community, and
their claims have been substantiated by studies showing
coexistence of species in spite of overlap greater than the
predicted 54%. These ideas will be explored in the following
chapter.

Chapter 5

Predation and species diversity

Many guilds and assemblages of potentially competing species appear to show a high degree of overlap in resource use without showing the differential partitioning of resources discussed in the previous chapter. These include parrot fish, tropical fruit and flower eating birds, tropical lacustrine fish, and many other herbivore guilds [165]. Other studies have looked for, but failed to find, interspecies influences and competition (e.g., guilds of folivorous (foliage-feeding) insects [124]). The existence of undiscovered resource dimensions provides one possible explanation. Otherwise, such coexistence can only occur if resources are naturally unlimited, or if potentially competing populations of coexisting species are kept below the carrying capacity of the environment through the action of some external force. In such cases interspecific competition will not be a principal mechanism determining species diversity and other factors must be examined. Physical disturbances offer one possibility (see Chapter 8), but predation has emerged as potentially the most important alternative community structuring mechanism to competition-mediated resource partitioning.

If some mortality factor acts most heavily on the top ranking competitor or the most common species (compensatory mortality), then competitive eliminations might be prevented indefinitely. This will allow greater niche overlap and hence more coexisting species [137, 165]. Theoretical studies have demonstrated that predation can act in this way and in so doing can affect local species diversity patterns.

5.1 Theoretical studies

The inclusion of density-dependent predation in MacArthur and Levin's model of minimum niche separation (Section 4.1) produced the following conclusions [165]: (a) minimum niche separation distance with a predator present is never larger than in its absence, (b) given this, as predation becomes more effective, so the minimum niche separation for invasion decreases. Similarly, frequency-dependent prey selection can produce stable equilibria in theoretical models of two competing prey species where none existed before [166]. The ability of a predator to show functional and numerical responses to changes in prey density would be necessary, but perhaps switching (disportionally attacking the most abundant prey) is a more important response in this context. Indeed switching has been found to stabilize one predator — *n* prey situations and is the only mechanism that can stabilize interactions where prey show complete niche overlap [166]. Generalist predators can act in this way. Prey preference by more specialist predators on the dominant competitor acts in the same way as predator switching and can stabilize theoretical interactions in models where no prey equilibrium existed before, as long as some niche separation occurs between prey species [166].

Both frequency-dependent predation and preference cause prey abundance distributions to become more even. This alone will increase measured diversity in a sample, but is also likely to lead to an increase in the number of prey species sampled [167].

The predation hypothesis thus suggests that selective predation on the dominant or most abundant species can maintain relatively high species diversity over ecological time. There is also the potential for a feedback system, where new invading predators are supported by new invading prey species [1]. Speciation over evolutionary time may also be possible through such processes as frequency-dependent selection, where predation pressure promotes diversity among prey by conferring an advantage on rare forms over the more common ones [168]. Predation may also have equally important effects on prey population dynamics and spatial distribution, which in turn affect community structure and function [302].

5.2 Plant—herbivore interactions

The preceding discussion is equally applicable to plant—herbivore interactions, and much of the evidence supporting the predation hypothesis has come from these trophic interactions. Some ecologists believe that herbivory plays a central role in the organization of plant assemblages, by altering the incidence of species of high competitive index in certain kinds of vegetation [196]. It is also considered to be an important selective force in the evolution of secondary plant chemicals and morphological traits in plants [169]. An alternative view is that the Earth is green because heavy grazing and depletion of plants is rare in natural systems, hence producers are not limited by herbivores but by competition for resources [170]. Harper believes that the reason why the animal is often not seen as an important factor in plant activity in natural situations, is that the most exciting phases of invasions and vigorous interactions from the early stages of producer/consumer interactions have been played out [75]. A case in point is the successful biological control of the once-dominant weed *Hypericum perforatum* in California by the beetle *Chrysolina quadrigemma*. Now, an ecologist would not recognize that *Chrysolina* is controlling *Hypericum* at its low density and hence maintaining the high species richness of rangeland plants [138].

5.2.1   *Terrestrial systems*
Large scale defoliation of high-ranking competitive plants can have the same effect on plant assemblages as consumption of the most competitive prey by predators does in animals. As mentioned above, such removal of plants is rare, and often unselective (e.g. locust devastation). However, herbivores can alter the relative fitness of competing plants through quite small amounts of grazing damage, thus indirectly bringing about community change [171, 172].

Grazing has unpredictable effects on above ground plant yield but may alter the competitive balance between the grazed plant and other species. Above ground grazing usually results in significant loss of below ground material as well [172], which must influence root competition and access to mineral resources and water. Below ground grazing can presumably have similar effects. Grazing can also lead to reduced seed number (especially under competition for light [169]) and reduced seed weight (e.g. annual crops, herbaceous perennials and trees [169, 173]). This can reduce competitive advantage in the following generation, as seedlings from small seeds may be at a disadvantage. The impact of grazing is also likely to be heightened or may only be effective, when the plant species is already under some other stress [172]. For example, a combined effect of rabbit grazing and low nutrient status probably maintains the high plant diversity on open grey dune systems [174].

The most complete studies of the role of herbivores on plant assemblages come from the literature on the biological control of weeds and from the management of introduced mammals on artificial grasslands (see Harper [75]). Selective grazing by sheep and goats, and less impressively by cows and horses, can increase plant diversity by decreasing dominant species. Indeed, Miles describes grasslands as the prime example of vegetation which can often be controlled or even created by grazing [4]. It is generally held that herbivory increases plant diversity by disturbing the process of successional advance, either by creating a microspatial mosaic of seral stages, or by preventing competitive domination by a few climax species [175]. Thus, grazing seems to affect successional relationships between the four most abundant vegetation types on acid soils in upland Britain [4]. Under low grazing pressure all communities revert to woodland, whereas under fairly high grazing, the trend is towards grassland (Fig. 5.1).

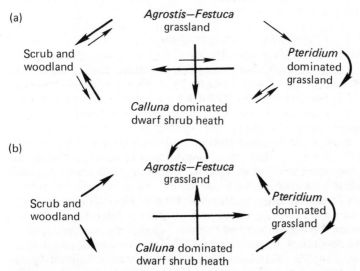

Fig. 5.1 Successional relationships between types of vegetation of well-drained acid soils in upland Britain, given (a) low or negligible, and (b) high grazing pressure. Large arrows represent normal sequences, small arrows less common ones. (From Miles [4].)

5.2.2   *Aquatic systems*

The effects of herbivores in some aquatic communities have
been demonstrated through exclusion and transplantation
experiments. Such experiments have shown, for example,
that sand plain algal species can competitively eliminate
species on the coral reef slope, but between-habitat diversity
is maintained, in the form of two distinct algal assemblages,
by herbivorous fish selectively feeding on sand plain species
that colonize the reef slope [176]. Total cover and species
composition of intertidal ephemeral algae can be influenced
by herbivorous dipteran larvae and crabs [177], but most
impressively by sea urchins. Exclusion of the urchin *Stongy-
locentrus purpuratus* leads to a lush plant assemblage dominated by large brown kelps, but selective grazing of the
superior benthic algae by urchins leads to a more diverse,
herbivore-resistant flora [178]. This situation is more complex where the predatory sea otter *Enhydra lutris* is present.
They effectively control urchin populations, which allows
kelp beds to develop. When otters are removed by hunting,
a 'cascade effect' occurs through the community as the
urchins now flourish and alter the composition of the plant
assemblage [179]. A similar 'cascade' is known for lobster-
influenced regions in Canada [178] and man-influenced shores
in Chile [306].

5.2.3   *Predation on plant propagules*

Instead of feeding directly on the plant itself, primary consumers may potentially alter or regulate the species composition of plant assemblages by feeding on seeds and fruit
[171]. Seed predation by ants and rodents, as demonstrated
by exclusion experiments, both enhance plant species diversity
in desert systems [64]. Ant feeding is frequency dependent,
whereas rodents selectively feed on large-seeded species
which tend to dominate plant biomass and out-compete
smaller-seeded species. Several authors have also suggested
that the action of seed predators could be responsible for
high tree diversity in tropical forests [171]. The seedlings
of some tropical species tend to grow under trees of other
species rather than their own, where the species specific
seed predators are commonest [180]. Provided there is some
ceiling on the abundance of each species then diversity can
be increased as a tree is more likely to be replaced by one of
another species when it dies [137]. Some circumstantial
evidence is used to support this hypothesis. Janzen [180]
suggests that the production of vast quantities of seeds at

one time can lead to predator satiation, which in turn will lead to progressively more interspecific competition and possible exclusion of species. Thus the decrease in species diversity on a gradient from the aseasonal tropics to seasonal tropics to temperate zones is explained by less-effective seed predation due to increasing predator satiation. This eventually leads to one species domination, as in the pine forests of the Russian Taiga, where the density of *Pinus cembra* is probably set by intraspecific competition. Further evidence comes from the unsuccessful attempts at growing the rubber tree in dense stands in its natural habitats of the Amazon basin whereas such plantations are successful in Malaya in the absence of natural enemies [10]. However, the generality of this mechanism has been questioned by the recent work of Connell (see Chapter 8).

5.2.4 Secondary consequences of grazing

These may also influence plant species diversity by creating or enhancing heterogeneity within the habitat. For example, a large faecal dropping can: (a) smother and exclude light from plants; (b) create local disturbances in nutrient relations; (c) change grazing patterns around the patch; and (d) create an island for colonization by new species [75]. Small animals can create similar islands of disturbance, such as rabbit latrines and mole hills. Intensive rabbit grazing may expose a larger proportion of the soil surface than normal to frost heaving and wind erosion, preventing or slowing stabilization of the soil surface thereby retarding succession [181]. The introduction of pathogens in combination with feeding may also alter the equilibrium between competing species of plant [171]. Finally, an increase in the rate of resource regeneration can also contribute to producer species diversity by alleviating resource limitation to some degree. Rapid recycling of plant nutrients through zooplankton feeding and defaecation is one possible example.

5.3 Predator—prey interactions

There is much evidence, especially from examples of biological control, to show that predators can reduce prey populations below the carrying capacity of the environment [10, 83]. Some parasites, in particular insect parisitoids, can act in the same way. This topic has been examined in detail elsewhere [182]. What concerns us here is not the impact of a predator on a prey population *per se*, but the effect that such an interaction might have on the rest of the community. The predator removes prey individuals and so releases resources which become available to other species. A parasite can have indirect effects, by reducing the ability of infected animals to compete successfully for available resources [183]. Plant parasites can act in the same way.

Patterns of guild structure in folivorous insects are similar to those in other groups, but appear to occur for reasons other than competition [124]. Early workers suggested that such herbivore populations must be predator limited, as lots of plants are still available and prey populations tend to increase in numbers when predators are removed [10, 170]. Such predation could lead to high species diversity amongst herbivores by the release of resources for additional species. However, food is not necessarily superabundant, especially for insect herbivores, due to secondary plant chemicals and low nutrient status [184]. Thus low population densities will not necessarily lead to high species richness. However, predators and parasitoids are recognized as one important factor, amongst several, which help explain species diversity of herbivorous insects [124, 184–186].

Most evidence supporting the predation hypothesis as an explanation of animal species diversity comes from studies on zooplankton and intertidal communities. Species richness of experimental communities containing planktonic crustacea feeding on phytoplankton can be dramatically altered by the introduction of fish [187]. Selective predation on *Ceriodaphnia*, a dominant herbivore, led to an increase in population sizes of most resident species, especially small cladocerans, and allowed the successful invasion of juveniles of three 'colonist' species. Selective predation by fish and salamanders on large zooplankton species is well documented [178, 188], and usually leads to an increase in smaller species number. Invertebrate predators, such as predatory mysid crustaceans, can act in the same way [189].

The most impressive studies have undoubtedly come from interidal communities, where space is the major limiting resource. In his classic study, Paine [190] experimentally excluded the starfish *Pisaster* from several sites in Mukkaw Bay, Washington. The result was a decline in diversity from 15 to 8 species and competitive domination of space by *Pisaster*'s preferred prey, the mussel *Mytilus californianus* (Fig. 5.2). This dramatic change occured in spite of the presence of other predator species. Removal of the starfish *Stichaster*

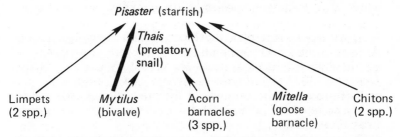

Fig. 5.2   The feeding relationships of the *Pisaster*-dominated subweb at Mukkaw Bay, Washington. The heavy line indicates a strong interaction. (After Paine [190].)

australis from the intertidal shoreline in New Zealand produced rapid domination by the mussel *Perna* and elimination of 19 other species [178]. In the same way, the crown-of-thorn starfish, *Acanthaster*, is thought to enhance diversity in coral reef communities on the Pacific coast of Central America by selective consumption of the competitively dominant coral *Porcillopora* [191]. Similarly, experimental exclusion of fish from artificial substrates led to domination by two species of colonial ascidians which are normally selectively removed, allowing colonization by a wide range of invertebrates [192]. In all these studies, selective predation reduces competition in the lower trophic levels by preventing the dominant competitors from monopolizing space, thus maintaining high diversity. Predators operating in this way are sometimes termed keystone predators [9].

5.4   Reduced community diversity through predation

Predation does not always lead to increased diversity amongst the lower trophic levels. Theoretically, although predation can reduce prey density, it does not necessarily mean that prey consumption of the resource is also reduced; a vital point if diversity is to increase. Indeed, alleviation of intraspecific competition may actually increase species activity and reproduction, leading to increased resource use. As mentioned earlier, herbivory may lead to substantial modifications in either direction of the root/shoot ratio and root activity. This may affect the extent of competition for light or between roots, and in turn result in high or low vegetation diversity [172]. The abundance of exploiters, feeding preferences and the effect of different frequencies of plants can also influence the effect of grazing [185]. For example, where the dominant species is unpalatable, grazing simply strengthens dominance and further reduces diversity [75].

Apart from some plant/herbivore interactions, low predation pressure seems to have no effect on community diversity, as found in intertidal systems for predatory gastropods like *Thais* [1,190, 302]. Likewise, where the preferred prey is rare or subdominant, removal of the predator has little impact on the community [178,189]. Such predators may be termed weak interactors.

Theoretically, in one predator two prey models, equivalent predation (no preference for either prey species) can make no difference to competitive coexistence among prey except on a local scale where a potentially stable equilibrium already exists [28, 166]. Diveristy can only be increased under such conditions if the competitively inferior species have higher population growth rates than the dominant ones. This may explain the situation where uniform grazing leads to increased species diversity of plants, where more 'r' selected species coexist with 'k' selected counterparts [166]. However, lack of preference coupled with intense predation pressure normally leads to reduced community diversity, as seen in mosquito larvae predation on a protozoan assemblage in pitcher plants [193], from the effects of the introduction of the catfish *Clarius gariepinus* on benthic invertebrate diversity in man-made ponds [188] and by *Littorina* grazing of intertidal microalgae [303].

Predation on one trophic level can also lead to 'cascade' effects in other levels, resulting in a reduced community diversity as a whole. This has already been seen in the sea otter — sea urchin — intertidal algae system, but a more dramatic example was documented after the introduction of the predatory cichlid, *Cichla orcellaris* (the peacock bass), into Gatun Lake in the Panama Canal. This predator is strictly piscivorous and it devastated the natural fish populations, including *Melaniris*, a key species in the lake's food web. As a consequence, there were drastic alterations in the producer, herbivore and tertiary consumer trophic levels as well (Fig. 5.3).

Finally, predation will only have an effect on community diversity if there are strong competitive linkages in the lower trophic level. Where no such interactions exist, it is unlikely that predation can be a major organizing force in that community. For example, exclusion of the conspicuous top predator of stream habitats, the trout (*Salmo trutta*), had no

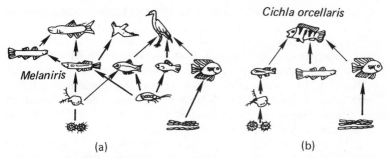

(a)   (b)

Fig. 5.3 Generalized foodwebs of common Gatun Lake populations before (a) and after (b) the introduction of the piscivorous *Cichla orcellaris*. (After Zaret and Paine [291].)

effect on the population sizes of the prey species, nor on community composition [194]. This is due to lack of competitive dominance amongst the prey, or to the top predator being a weak interactor for some reason.

5.5 Conclusion

The predation hypothesis appears to be substantiated only for a limited range of examples and communities, mainly involving two-dimensional partitioning of space. This observation does not support the strong statements made that 'predation is as important in controlling species richness and organizing communities, as competition (e.g. [83]). Indeed attempts to apply the predation hypothesis to marine benthic communities in the three-dimensional soft sediments have been unsuccessful, and from experimental studies, predation actually seems to lead to a lower species diversity than is otherwise possible [304]. In reality, a number of criteria must be met before predation can lead to high community diversity.

(a) Predator preference is essential.
(b) Strong, competitive, asymmetric cross-linkages among the lower trophic levels within the food web is equally important. The influence of top predators in intertidal communities is due to the predator playing a key role in controlling its preferred prey, which in turn is the competitively dominant species of the lower trophic level (i.e. both are strong interactors). These cross-linkages amongst prey species then translate changes in predator abundance into the allocation of the limiting

resource of space, and account for the dramatic cascade effects which can occur when the community is disturbed [178].

(c) The above presupposes that resources are freed through the influence of the predator and can support new species or larger populations of species already present.

(d) Predation intensity is important. Only if the feedback force between predator and the preferred, dominant, prey is great enough can predation maintain community diversity in the face of perturbations [75, 167]. Low predation intensity usually has no effect unless occurring at the same time as other forms of stress. For example, a combination of nematode feeding and interspecific competition is sufficient to decrease the competitive advantage of oats and allow existence of barley (see [172]).

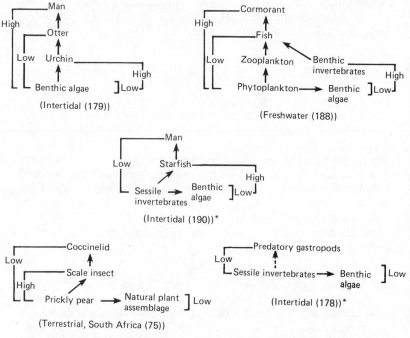

Fig. 5.4 Changes in diversity and complexity (Low or High) with different numbers in interacting levels. Arrows indicate direction of interactions. Strong interaction ↑, weak interaction ↑. * Sessile invertebrates and benthic algae are put on the same level as they compete for space. References are indicated in the diagram.

Similarly, folivorous insect diversity has been explained by a combination of predation, parasitism, plant distribution, plant chemistry and vagaries of the weather [124]. However, intense predation can lead to a decrease in species richness, and on the basis of some limited evidence from herbivore/plant interactions, species richness is maximal under moderate levels of grazing (i.e. a humped relationship between predation intensity and species richness [295, 305]).

(e) To maintain high species diversity, the controlling agent must not only become the main regulator of the dominant prey species, but must itself be regulated by the availability of the prey and not limited by its own predators. If such limitation does occur, predation at this higher level will lead to reduced diversity among the lower levels through cascade effects. Successful biological control of weeds also requires the divorce of the herbivore from its own predators [75]. A pattern seems to emerge, where an even number of strongly interacting levels leads to greater diversity and complexity, but an odd number to dominance and simplicity (Fig. 5.4). The generality of this pattern has yet to be established.

Chapter 6

Competition
and predation:
complementarity
of the hypotheses

6.1 The controversy

The importance of adaptation to physical factors in the control of species distributions and abundances over broad geographical areas is unquestionable. However, in most cases, the proximate limits to species ranges are set more by biotic factors. Traditional views are that competition between species on the same trophic level is the dominating biotic influence, but until fairly recently, much of the evidence from nature was circumstantial, based on observed discontinuities of distributions and differences between allopatric and sympatric resource use. The significance of such evidence is open to debate, as the observed differences in cases of niche shift and character displacement for example, might arise because the areas compared differed in some way other than the presence and absence of related species. This led some authors to challenge the dominant role of competition in the regulation of species diversity (e.g. [83]) and to propose that biotic interactions between trophic levels (notably predation and herbivory) provide alternative regulatory mechanisms. However, the widespread demonstrations of niche shift, the success of manipulative experiments and the large body of data collected over the past few years that demonstrate competitive effects both quantitatively and qualitatively, clearly indicate that such challenges are generally weak. (See Roughgarden [301] for similar philosophical arguments.)

In some cases, where competition has not been found to occur regularly on a day to day basis, it could still be a significant force. Active avoidance of interspecific competition in itself implies that competition has occurred at some time in the past, so the species have adapted to each others' presence and are kept apart in the same way [9, 44]. Inefficient competitors should not, then, be normally observable, as they will be quickly eliminated by competitive exclusion, especially amongst short lived, small organisms like insects and annual plants. This explanation is acceptable only if the competitors coevolved. Connell has challenged this by suggesting that competitors are not dependent on each other and so need not consistently co-occur or coevolve, especially as species number and relative abundances increase [195]. Thus any differences are more likely to have evolved in isolation. Coexistence between species is possible because of such differences being already present in their ecologies. These views have been supported in the case of folivorous insects [124]. This argument is also weak [e.g. 301]. It is just as difficult to show that two species coevolved as it is to show that they did not, so this approach is unhelpful. Also, the importance of competition in island communities cannot be disputed in this way, as the mainland source of colonists is usually known, and the ecologies of island and mainland populations of species can be compared directly.

In contrast, it is argued that as predators are dependent on their prey, they must coevolve, and so predation should be a dominant factor in community organization as outlined in the predation hypothesis [195]. However, predator effects are much more direct than competition, and weak or susceptible phenotypes and genotypes are likely to be very rapidly eliminated from a species population. This will result in rapid evolutionary changes leading to anti-predator adaptations allowing the prey to coexist with the predator. Adaptations associated with morphological (spines, stinging cells, camouflage), chemical (nasty tastes, poisons, tannins, alkaloids etc.) and behavioural (aggressive defence, size, use of refuges) characters are widespread and important features of predator/prey interactions. The predation hypothesis is relevant when the level of predation is great enough significantly to reduce population levels of the dominant competitor. Anti-predator adaptations, therefore, must ameliorate this impact and so reduce the effectiveness of predator influence on community structure below that envisaged by some authors. (See also [302].)

6.2   Interactions between competition and predation

To demonstrate that species differences have been promoted by competition it must be shown that these differences are greater than other simple factors dictate. For instance, phytophagous insects might differ in ecology in order to avoid their neighbour's predators and parasites [124]. Frequency-dependent selection can also lead to phenotypic differences in appearance, and rare prey escaping predation. Over evolutionary time, this might lead to speciation and resource partitioning. There might also be strong selective pressure on a consumer to gain energy quickly while feeding, not only from a competitive point of view, but also to reduce the chance of its being eaten itself at that time. Predators may thus reinforce selection for feeding efficiency, including resource specialization. It can then become difficult to disentangle resource partitioning differences from differences related to predator avoidance [136].

This interaction of predation and competition has been demonstrated theoretically [165]. The minimum niche separation distance between two species on a resource continuum is determined jointly by the level of predator pressure and the kurtosis of the competition functions (a direct reflection of the shape of the resource utilization curves). Niches can become closer as predation pressure increases and/or competition functions become more leptokurtic (Fig. 8.1(e)).

Grazing may only influence plants when superimposed on some other stressful interaction (Chapter 5), and the outcome of competition between plant species in some circumstances is only explicable in terms of invertebrate grazing occurring at the same time [172, 175]. In the same way, cascade effects and predator/herbivore impact in intertidal communities can only be explained by the presence of strong competitive cross links in the lower trophic levels.

6.3   Classification scheme

The above discussion suggests that some complementarity exists between the competition and predation hypotheses, and there have been several attempts to generalize those situations where one or other mechanism should predominate.

6.3.1   The physical environment

Connell [195] suggests that in very harsh conditions, populations are decreased below levels at which they compete by physical rarefaction, and in benign conditions by predation. In intermediate, moderately harsh environments, where mortality from direct physical stresses and from natural enemies should decline, populations are more likely to reach levels at which they will compete and so competition becomes the dominant organizing force (Fig. 6.1). Grime [196] suggests something similar to explain plant diversity, where increased environmental stress (physical influences, or herbivory) reduces the vigour of competitively dominant plant species and increases species diversity. The supporting evidence is somewhat sketchy. There are some examples where plant interactions are alleviated by weather [195] and pollution or fire [172]. There appears to be a negative correlation between the proportion of alkaloid-producing trees and latitude, whereas the efficiency of wood production tends to increase at higher latitudes [176]. This suggests that herbivore resistance is most important in the benign tropics, whereas competition gradually dominates with increasing latitude. The proportion of predatory species is also known to be relatively greater in certain diverse situations [190].

This scheme ignores the fact that in benign environments competition among predator species will probably be increased, which could reduce their impact on prey populations. On the basis of energetics, a greater diversity of prey species can maintain more predators. This could mean that the relative influence of predation on the competitively dominant species is no greater than on any of the other prey species. Predation itself can also be moderated in complex environ-

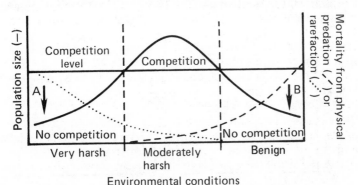

Fig. 6.1 Connell's [195] classification of community organization mechanisms based on environmental conditions. A — populations depressed by physical rarefaction; B — populations depressed by effective predation.

ments, the benign tropics being an obvious environment. In simple environments, predators will find it easy to search, so could doubtless exert control over species diversity [28]. With increased complexity, prey species will be hard to find due to the provision of refuges and there is likely to be a decline in predator foraging efficiency. This can be shown by a functional response which describes the change in the numbers of prey attacked in a fixed period of time by an individual predator as a result of change in prey density. In simple environments, a Type II functional response is normally found [52], described by a negatively accelerating curve rising to an upper plateau (Fig. 6.2(a)). Predators in complex environments often show a Type III response [197], which is described by a sigmoid curve (Fig. 6.2(b)). Experimental manipulation has shown that increased structural complexity of the environment can reduce the predation impact on a competitively dominant member of sessile marine invertebrate assemblages [192] or reduce predation risk to desert rodents [198]. Similarly, a gradual increase in environmental complexity can alter the functional response of net-spinning predatory caddis larvae from a Type II to a Type III ([199], personal observation). Structurally complex localities with many microhabitats are more likely to allow coexistence of many more species through competition-induced habitat selection than through an increase in predator impact.

6.3.2   *Trophic status*
A more satisfactory approach to classification is based on trophic status and size [1, 136, 170]. Competition can regulate the number of species in a guild only when popula-

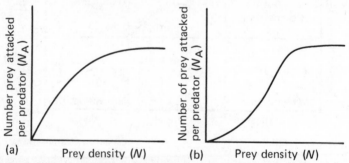

Fig 6.2   Two types of predator functional response showing the changes in numbers of prey attacked per unit time by a single predator as the initial prey density is varied. (a) Type II response, (b) Type III response.

tions are at or near the environmental carrying capacity. Menge and Sutherland [1] suggest that this should apply to animals of high trophic status that are not so affected by environmental fluctuations (i.e. homeotherms), show parental care, and are unaffected by predation. Large predatory animals thus provide classic cases of resource partitioning [136]. Conversely, guilds of lower trophic status, such as herbivores and small animals are more likely to be regulated by predation (Fig. 6.3(a)). There is some support for the trend that smaller animals compete less than larger ones [296, 297].

Alternatively, Hairston *el al.* [170] suggest that competition should be found in carnivores, producers and decomposers, but not in strict herbivores (i.e. the competition curve in Fig. 6.3(a) should be U-shaped). This trend has been demonstrated for both freshwater and terrestrial species [296], but only weakly for marine species (on which the other model is based).

It has also been suggested that in communities with few trophic levels, competition will be the most important overall organizing factor [1] but as the number of trophic levels increases and the number of species increase, predation becomes relatively more important (Fig. 6.3(b)). This is much less acceptable as a general rule, not only from consideration of the previous section, but also because predation can increase or reduce community diversity depending on the number of interacting trophic levels (see Chapter 5).

6.4 Conclusion

Although the controversy over the relative importance of competition and predation in animal assemblages has apparently not been resolved, some degree of consensus seems to

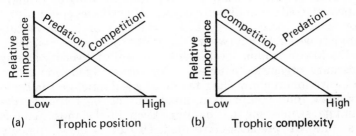

Fig. 6.3 Qualitative models of the relative within-community (a) and between-community (b) importance of interspecific competition and predation in organizing and maintenance of diversity in communities. (After Menge and Sutherland [1].)

have been reached among plant ecologists. Vertebrate grazers can control the physiognomy of vegetation in some cases, but for most grazers, especially insects, the influence seems to be restricted to effects on the population of single species [4, 170]. Most vegetation is, therefore, probably resource limited [75, 295]. Interspecific competition is then likely to be the main driving force determining the distribution and abundance of plants, but the direction this takes can be determined to some extent by a herbivore [172]. Predators alleviate resource limitation by reducing the activity of the dominant exploiter of resources. A point not often appreciated, is that if such a predator-controlled community is not yet saturated with species, it can still become saturated with individuals. Thus, many of the resources freed by reduction of the dominant competitor will be taken up by these other species. So, despite the predator activity, resources will still be saturated, competition between coexisting species will continue, and it is this competition which still controls the diversity, relative abundances and resource use of species at this point. In some circumstances, predation may alter the number of coexisting species, but such communities are not continually open to new species. A limit is reached when resources are again saturated. Even if predators keep all species low by rapid switching between prey, the freed resources will allow other invasions until resources are again saturated. Resource use and species abundance will still be controlled by competition.

Emphasis has been placed on competition as the organizing factor of communities both because of its effects on niche size and because of the way that Hutchinson defined the niche [195]. This has to some extent been misleading, as predation does offer a potential alternative mechanism. However, predation cannot be regarded as being of primary importance either in directly determining species composition (for which there is very little evidence) or in preventing competitive exclusion (where the evidence is largely restricted to the lowest trophic levels and to control of diversity among sedentary organisms competing for space). Predation might be a proximate controlling factor of species richness in some instances, but competition must still be operating as the ultimate structuring mechanism. Even in the most famous examples of predator-controlled species richness, namely intertidal communities, the predation effect is simply superimposed on a bed of competitive interactions among the prey.

Chapter 7

Saturation of communities

Increasing niche dimensionality, differential niche overlap and occasionally predation, all lead to decreased competition and hence may allow more species to coexist than would otherwise be possible. Whittaker [11] has expanded on this idea by suggesting that new species added to a community are themselves resources and increase the breadth of the resource spectrum for other species. The addition of species to communities is thus a self-augmenting evolutionary process — diversity begats diversity. Support for this view is taken from traditional ideas that diversity leads to stability and hence to reduced extinction (but see Chapter 10). The rate of species diversity is, therefore, endowed with a positive feedback; the more species, the lower the extinction rate and the faster diversity grows.

The argument is logical, but there must be some limit given the worlds energetic and nutrient limitations, and because each species must contain at least one individual [200]. If most species compete significantly, or have competed in the past, with similar species, and if there are limits on total niche space, niche width and tolerable overlap, then one would predict a limit to the number of species an area could maintain. As competition appears to be the ultimate controlling factor of species richness, one would expect to find communities which had achieved such saturation and a steady-state diversity. Species saturation has also been predicted on the basis of theory originated to describe island species diversity but which has since been extended to continental 'islands' and mainland areas.

Is there a maximum number of different species that can coexist within an ecological system? Limited evidence suggests that some communities and portions of others may indeed be saturated with species.

7.1　True islands

These offer the most tractable systems with which to begin investigating the regulation of species diversity in communities, and have thus been the subject of intensive study. This has been based around the Equilibrium Model of MacArthur and Wilson (see [28]), which in essence, predicts that the biota on an island is under a dynamic equilibrium between immigration from the source pool area (which decreases with increasing diversity through the exhaustion of new colonist species) and extinction of colonists (which increases as more and more species are packed on). If nett immigration and extinction rates are plotted against species number, the point at which the two curves cross describes the equilibrium species number (Fig. 7.1(a)). In reality, the single sharp curves normally portrayed should be viewed as blurs, to take account of all the separate curves possible for different orders of arrival and extinction of species. This would form a relatively large area of intersection, thus providing the variation around the equilibrium species number that is often found [28]. Competition will cause a decrease in successful immigration and an increase in extinction with increasing species number [201].

The occurrence of an equilibrium is now generally accepted. Species numbers have been shown to increase to their appropriate equilibrium where a natural catastrophe or experimental manipulation reduces the flora and fauna [201–203] and to decrease to equilibrium when islands are evolved from land bridges or are reduced in size when sea levels rise [203]. There is some debate as to whether the equilibrium is dynamic and an alternative model of island diversity suggests that certain species are permanently resident (S^*), with zero extinction rates, whilst other immigrants fail to breed, and so have an extinction rate of infinity (Fig. 7.1(b)). There is still an equilibrium, but no turnover of species. There is some evidence of a dynamic species turnover at equilibrium amongst birds [28, 203, 207] and arthropods [201], so a combination of the two models may be most appropriate, where there is a permanent core of resident species but turnover of others (Fig. 7.1(c)). There also appears to be a decrease in the rela-

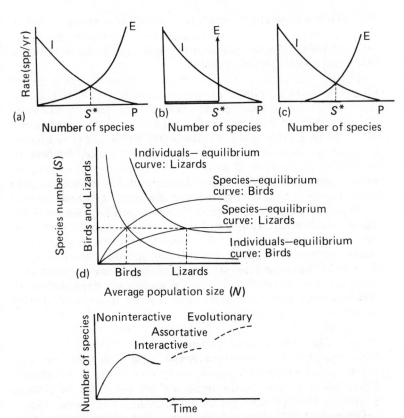

Fig. 7.1 Models of species colonization on islands. (a) Equilibrium model of MacArthur and Wilson [28]. (b) Permanent resident model (After Diamond and May [203].) (c) A combination model of the previous two. (After Diamond and May [203].) (d) A model predicting the number of species (S) and average population size (N) for lizards and birds of small islands. (After Schoener and Schoener [204].) (e) Phases of species colonization (After Krebs [205].) S* — equilibrium species number; P — size of species pool; I — immigration curve; E — extinction curve.

tive turnover (% per year) from 'lower' to 'higher' organisms with a roughly linear relationship between log turnover and log generation time [308]. Further elaboration of the equilibrium model has introduced a third axis, the average population size (N). Immigration and extinction are related in the same way to species number (S) but immigration is unrelated to N and extinction decreases with increasing N [204]. The resulting intersecting surfaces give a species equilibrium curve if projected onto the *S—N* plane, representing those combinations of *S* and *N* at which equilibrium occurs on the island (Fig. 7.1(d)). These predictions have been supported for birds and lizards on small islands.

During the process of colonization, islands may actually go through several phases of species equilibria [28, 201, 205]. An initial stable number of species may rapidly colonize the island producing a non-interactive or quasi-equilibrium (Fig. 7.1(e)). This may be followed by a slight reduction in species number during an interactive phase, leading to sequences of coadapted groups forming interactive equilibria. (Such changes have been demonstrated experimentally with sessile aquatic organisms colonizing artificial islands [206, 207].) Further coadaptation will produce successively larger assortative equilibria, with lower turnover rates and over evolutionary time still higher equilibria as species in more highly coadapted complexes should evolve smaller niches. A limit will be set to species packing by increased probabilities of extinction as population sizes fall with further narrowing of niches and from a limiting similarity set by competitive interactions.

7.1.1   *The species—area relationship*
The equilibrium theory predicted that isolation and area are the major factors controlling species richness on islands. Large islands close to the mainland are predicted to possess a greater species richness at equilibrium than more distant small ones [28]. The species—area relationship is the most important in the present context, and the literature demonstrating its generality is vast and still growing. The relationship is found on many natural islands [85], and has been demonstrated experimentally with artificial islands [206, 207]. A recent review has collated over one hundred examples [208].

For both plants and animals, and within a variety of taxonomic groups, the relationship between island size (*A*) and species number (*S*) is given by the power function

$$S = CA^z$$

C is a constant and on a log *S*/log *A* plot, the slope of the regression line, *Z*, most often falls within the range of 0.18—0.35 [203]. Thus a tenfold decrease in area results in half the equilibrium number of species. (Other mathematical descriptions of the relationship are possible [208].) An independent explanation for this particular relationship has been proposed by May [209]. A species—area relationship based on the log-normal distribution of species abundance (see Chapter 9) and proportionality between the number of individuals and island area [114], is well approximated by the above power function with a slope $Z = 0.25$. Statistical artifacts related to log transformations and linear regression can cause most Z-values to fall within the circumscribed range 0.2—0.4 [208], but more detailed theoretical analysis suggests that such values reflect real regularities in the shape of the lognormal distribution [210]. There is obviously some dispute on statistical grounds and unfortunately most studies use the linear regression technique. If iterative non-linear regression is used to fit the power function, there are no statistical artefacts and in a review of 36 species—area relationships [84], 27 showed Z valves of between 0.15 and 0.39. This gives greater credence to the generality of the relationship and provides support for May's theory. Even so, many workers still maintain that Z has no unique biological significance [309, 310].

7.1.2 Explanations of the species—area relationship
There is no doubt that the number of species on an island increases with area, and several hypotheses have been advanced to account for this.

(a) Species—area relationships are solely a sampling phenomenon, where larger areas receive effectively larger samples than smaller ones, so contain more species. This has been suggested for lichen species on 'islands' of different sized boulders in subtidal areas [211], but denies the importance of population processes and habitat diversity.

(b) The habitat diversity hypothesis proposes that as the area sampled increases, so new habitats are encountered. Small islands entirely lack certain habitats such as lakes, streams and grassland [213], and complex habitats have lower extinctions curves [28]. Thus increasing habitat diversity will lead to increasing species number on islands, and several studies have demonstrated such a positive correlation [208, 214].

(c) The area *per se* hypothesis is predicted from the equilibrium theory. Larger islands will have a higher immigration rate than small ones as they present a larger target [28], and will also support larger species populations which have a smaller extinction rate. Species requiring a minimum territory size or viable population may also be restricted to larger islands [204, 213]. Thus larger islands will possess a greater equilibrium number of species simply because they are large.

Each hypothesis is probably important in determining the correlation between species number and area in one or another species assemblage, but due to the interaction between area and habitat diversity, it is often difficult to assess the proportional contribution in any particular study. Where habit diversity is controlled, and area altered experimentally, as in Simberloff's studies on mangrove islands [201] and those of Rey on saltmarsh islands [312], or naturally [215] as in Terburgh's study on island birds, area clearly has an independent effect on species number. Multiple regression can also be used to disentangle the effects of various factors, and with birds [28], birds and lizards [204] and bats [313], island area is a better predictor of species number than habitat diversity.

The important point is that irrespective of the controlling mechanism or the statistics, the existence of the species—area relationship itself implies that some maximum number of species can be packed into an area. If there was no saturation limit, such a clear positive relationship would not occur. Such island saturation has been directly demonstrated for bird assemblages (Fig. 7.2) where it is suggested that a limit to tolerence of competition imposes a ceiling on diversity [215].

7.2 Habitat islands

Any patch of habitat that is isolated from similar habitats by different, relatively inhospitable terraine (for the organisms concerned) may be considered to be an island. Such habitat islands include freshwater bodies, desert oases, woods, caves and montane areas. Consequently, models derived for island biology should also be applicable to small-scale local systems on continents. Indeed these habitat islands show equilibrium species diversity and similar species—area realtionships to the true island situation. Examples include avifauna and flora in Paramo vegetation in the Andes, montane mammal assemblages in cool boreal habitats of the Great Basin, USA, and limestone cave fauna. Further details are given by Gorman [61].

Most continental habitats also form complex mosaics, each habitat patch being surrounded by other terrestrial com-

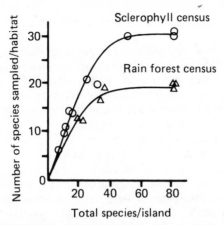

Fig. 7.2 The relationship between total number of breeding land bird species per island and the number sampled in two habitat types at 17 sites on 12 Greater and Lesser Antillean islands. (After Terborgh and Faaborg [215].)

munities, rather than a vacuum, as in true islands. It is likely that immigration into these mainland islands will be more frequent and successful, if species in surrounding habitats are not too different. This can lead to continual exchange of species with surrounding areas, or spillover of organisms from adjacent habitats adding to the habitat specialists [28]. Either of these might obscure any insularity. However, a local equilibrium or saturation of species in such habitat patches might be expected based on immigration from other patches and competitive exclusion proportional to the square of the number of species present, as suggested for plant assemblages [211, 212].

7.3 Host plant islands

Plants, at the individual, population, or species levels of organization, can be thought of as host islands for phytophagous insects and plant parasites, and as such, Island Biogeographical Theory might well be applicable. This approach has been used in several studies as a means of explaining species diversity patterns of insects and fungi [124, 216—218]. In practically all cases, a species—area or species—range relationship has been found, where large clumps of a host plant, or plants with the greatest geographical range, support more species of phytophagous insects and parasites than plant species with small stands or ranges. The Z values tend to be higher than for true islands (see [265]). Some examples are given in Fig. 7.3. The debate is continuing over both the suitability of treating hosts as islands and over some data used to demonstrate species—area relationships (see [219, 314] and replies [220, 221, 315]). However, the relationship is empirical, widespread and found under so many different conditions that rejection on the basis of suitability of a few examples is invalid.

Area or range accounts for 20—90% of variation in species richness in these studies [217, 218]. This is either due to area *per se*, or because increased environmental diversity is correlated with a wide-ranging plant (i.e. a greater variety of climates or other plant species is encompassed). Although for some taxa such as leafminers [222], species—area relationships can be predicted on the basis of immigration (oviposition) and extinction (mortality), the demonstration of species—area relationship does not by itself necessarily imply such a dynamic equilibrium, as described for true islands [217, 218]. However, such relationships again show that a limit to species number occurs, i.e. a form of species saturation. Just as defaunated or new islands acquire a fauna commensurate with isolation and area, so introduced plants acquire insect faunas over ecological time commensurate with the size of the plants' geographical ranges [124] and their taxonomic relatedness to native plants. In addition, beyond 100—1000 years, time has no clear effect on the total number of species

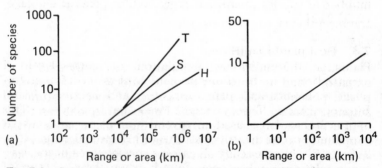

Fig. 7.3 Species/host plant island area relationships. (a) Insect parasites of British trees (T), shrubs (S) and herbs (H). (After Strong and Levin [217].) (b) Leafminers and British trees. (After Claridge and Wilson [218].)

colonizing the introduced plant species so demonstrating the saturation of such assemblages. Species richness of plant parasites show a similar rapid saturation within short periods of evolutionary time [217]. As host plant ranges and habitats contract or expand, extinctions or colonizations of insect species occur as expected in true island situations [221].

It has been suggested that competition, which contributes to increased extinction and the dynamicity of the equilibrium model for islands, is rare among phytophagous insects (see Chapters 3 and 4). What then prevents the continuous rise in species numbers as predicted by Whittaker [11]? One plausible hypothesis suggests that introduced plants initially draw colonists (depending partly on an ecological and bio-chemical match between the native and introduced plant) from a pool of native insect species [124, 316]. Thus rapid accumulation to an asymptotic species number occurs through rapid exhaustion of the species that have a high probability of colonizing the plant. This is something like a non-interactive or quasi-equilibrium, set at a level above that which competition would allow, through the main-tenance of low population sizes, possibly by plant defences and predators. An assortative or evolutionary equilibrium species number might develop through colonization of new species in evolutionary time. A true immigration (speciation)—extinction equilibrium might then emerge. The difference between host plant islands and true islands, however, will be in the coevolving nature of the island/colonist (producer/consumer) interaction, an interaction that does not occur with non-living true or habitat islands. It is not surprising that there are some difficulties in applying island biogeo-graphical theory to such host islands.

7.4 Continental saturation

Diversity patterns on continents are not as easily studied as on islands, and are thus not as well understood. However, it is still possible to identify some level of species saturation amongst mainland communities. For example, one can think of the process of succession (following a disturbance in the community) as leading to some form of steady-state diversity, and there is repeatable convergence on the same climax community from any of many different starting points [223].

7.4.1 *Mainland species—area relationships*
As expected, the number of species increases with increasing

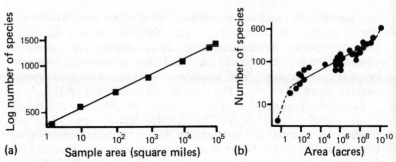

Fig. 7.4   Mainland species—area relationships. (a) Species of flowering plants in sample areas of England (After Gorman [61].) (b) North American birds. (After Preston [293].)

area of search on mainlands, but rather than a simple linear relationship, a decelerating increase in species number occurs so that above a certain area, few new species are found (Fig. 1.1). This is often recognized as the minimum area that would include a community [6].

The mathematics of the relationship are similar to those for islands, but plots of log species number against log habitat size yield lower Z values, in the range $0.12-0.17$ (Fig. 7.4). This indicates that small areas of mainland contain almost as many species as large areas. These arbitrary mainland areas are continually exchanging species with other surrounding areas and so do not suffer from the low levels of immigration found through isolation of islands. An associated low extinction rate might also contribute to the lower Z values [211]. The area effect *per se* is thus less important, and the relationships are more likely to reflect changes in habitat diversity.

In these situations, one confronts a problem of scale and different measures of species diversity [11]. A small patch of habitat supports a number of species, which is described as α or within-habitat diversity. When comparisons are made between habitats or along environment gradients, changes in species composition occur. The degree or rate of change is termed β or between-habitat diversity. The following equation can provide a simple estimate:

$$\beta = \frac{Sc}{\bar{S}}$$

where Sc is the total number of species occurring along a transect or in a set of samples (counting each species only

once) and $\bar{S}$ is the average number of species per individual
sample. For a single sample, $\beta = 1$, whereas for two samples
with no species in common, $\beta = 2$. However, the species—
area relationship on mainlands, and the asymptote reached
by such relationships, still implies some limit to species richness.

7.4.2   Models of mainland saturation

Mainland saturation has been predicted on theoretical grounds,
mirroring the development of the equilibrium model for
islands but with speciation replacing the immigration curve.
MacArthur [224] suggested that (a) speciation will increase
at a decelerating rate with diversity (more species means
greater opportunity for new species, but leads to less opportunity for success), and (b) extinction rates will increase
at an increasing rate (through competition and reduced population sizes as the community suffers increasing packing).
This should lead to a steady-state diversity (Fig. 7.5(a)).
Alternatively, conventional wisdom suggests that speciation
rates occur faster when there are more unfilled niches, leading
to adaptive radiation. It has also been suggested that most
geographical isolates (which normally lead to new species)
fail to produce new species because they do not find open
niches [225]. If this is so, Fig. 7.5(a) is wrong, and speciation rates may well decline with diversity (Fig. 7.5(b)). Two
types of speciation may account for these differences; (a)
competitive speciation, which is high when the biota is
depauperate as there are many opportunities for speciation
through competitive interaction, but it is a decreasing function of diveristy; (b) geographical speciation, the more species
there are, the greater the number of isolates formed by a

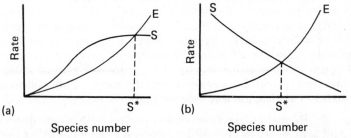

Fig. 7.5   Two models of mainland species saturation. (a) MacArthur's
model [28]. (b) An alternative model based on Mayr [225]. S — speciation curve; E — extinction curve; S* — equilibrium species number.

geographical barrier, and the more quickly new species are formed. This is an increasing function of diversity [226]. Either way, diversity is predicted to reach a steady-state, as opposed to the indefinite increase proposed by Whittaker.

7.4.3   Evidence of mainland saturation

There are several convincing examples demonstrating that steady-state diversities have been reached in some communities and portions of others. The comparability in α diversity of bird assemblages over similar habitat gradients in three Mediterranean-type localities (Fig. 7.6(a)) is presumably produced by natural selection via competitive interactions which determine the same degree of limiting similarity of coexisting species [42]. A similar study has demonstrated a similar strong correlation between bird species diversity and foliage height diversity on two continents (Fig. 7.6(b)), suggesting that these avifaunas are saturated [28]. For insects, lizards and some plants, such neat convergences do not occur [9], suggesting that these groups may not always be saturated with species. Some desert rodent assemblages appear to be saturated, as species similar in size replace each other on different dunes, and the most diverse assemblage consists of

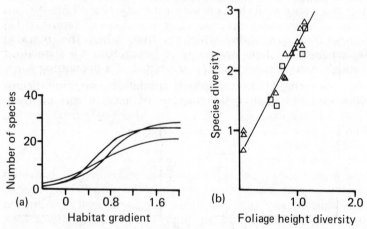

Fig. 7.6   Examples of species saturation. (a) Comparability in α diversity of bird assemblages in three Mediterranean-type localities (After Cody [42].) (b) Comparability in the relationship between bird species diveristy and foliage height diversity in Australia (□) and North America (△). (After MacArthur [28].)

five or six species [61]. Similar stability can be found in soil mite assemblages in spite of large changes in species composition through the year [154]. Stream invertebrate assemblages also demonstrate a constancy in α diversity despite a high β diversity with geography and time [227].

7.4.4 Historical evidence

Recent analyses of species richness throughout the Phanerozoic times (Cambrian to recent), indicate that both marine invertebrates and terrestrial plants have spent at least 350 million years at equilibrium diversity, despite a high species turnover [226]. For example, studies of cores from the Nicolet Valley, Quebec, suggest that marine invertebrate diversity rose sharply from Ordovician to Silurian times from a few to 35 species (through competitive speciation). This was then maintained for at least 50—70% of the 5 million year record examined, with a turnover of 20% species per million years (produced by geographical speciation).

If one can accept generic diversity as a fair reflection of species diversity, then the number of genera of North American land mammals provides more evidence of species saturation. There has been a balance between extinction and orgination for at least 10 million years with only occasional non-equilibrium episodes caused by habitat fractionation, lowered sea levels and other global environmental changes [201]. Lastly, if species diversity in two neighbouring biogeographical provinces are at steady state and become joined by geological processes, one would predict an increase in extinction rates [200]. At the family level, such waves of extinction took place in mammals at the recent union of North and South America and catastrophic extinctions fairly typical of many fossil animal families during the Permo-Triassic period coincided with the formation of Pangea.

7.5 Conclusion

In contrast to Whittaker's ideas of an ever-increasing species diversity, one must conclude that the number of species coexisting at any point in space (i.e. α diversity) is limited in ecological time, and in many communities, over long periods of evolutionary time. As one examines species diversity on a larger scale, such a limit becomes less obvious because of horizontal replacement of species (β diversity). This can allow the coexistence within an area of more species than can actually share the use of a common point in space.

Such between-habitat diversity may show some continual rise in evolutionary time, but a limit on the horizontal turnover of species within an area will occur (at a viable minimum area or population size for species survival), as shown by the type of relationship in Figs 1.1 and 7.4. Expanding the area of search still further, to encompass similar habitats in different zoogeographical regions, we find further turnover of species due to historical reasons (termed γ diversity). Logic again decrees some limit, but because of the continuing changes occuring through continental drift and evolution, such a limit will be hard to define.

Species saturation does exist within habitats, and appears to be due to two possible mechanisms. Firstly a non-interactive one, determined only by properties intrinsic to the habitat and the species in question (as seen in phytophagous insect assemblages for example). Secondly, an interactive mechanism, where species diversity is restricted by direct or diffuse competition within the habitat. Through evolutionary time, the non-interactive steady state might then gradually change to an interactive one.

Chapter 8

Species diversity
trends — theories
and hypotheses

We are now in a better position to try and explain the species diversity trends described in Chapter 1. If communities can become saturated, then species diversity differences may be simply related to the age of the community and the degree of saturation, i.e. the extent to which all available resources are exploited by as many species as possible. However, saturated communities can also differ in species diversity. MacArthur [28] derived a simple equation to approximate species diversity of animals (Ds) using a common resource

$$Ds = \frac{Dr}{Du} (1 + C\bar{\alpha})$$

where Dr describes the diversity of resources used by the entire community, Du describes the niche width of each species (assumed identical), C is the number of potential competitors or neighbours in niche space — an expression of the dimensionality of the habitat, $\bar{\alpha}$ is the mean competitive coefficient or mean niche overlap. Although little used, this formulation provides a useful basis for the following discussion, and predicts that saturated communities can differ in species diversity in three, potentially interacting ways.

(a) In diversity of available resources (extent of total niche hypervolume space). An area with a greater range of available resources (larger Dr and C) can maintain more

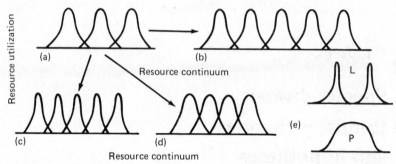

Resource utilization

Fig. 8.1 Species can be added to a community (a) by several mechanisms: increasing the resource base (b); increase in species specialization (c); increase in species overlap (d); changing niche shape from platykurtic, P, to leptokurtic, L (e).

niches and hence support more species than an area offering a smaller range (Fig. 8.1(b)).

(b) In niche width or degree of specialization of component species (Du). If species exploit smaller portions of the total niche hypervolume space (i.e. have smaller niches) the number of coexisting species can increase (Fig. 8.1(c)).

(c) In the average degree of niche overlap between species ($\bar{\alpha}$). Greater overlap of species (smaller exclusive niches) leads to greater diversity even if two communities are similar in both (a) and (b) (Fig. 8.1(d)).

Returning to the simple 'ball and box' model of the community used earlier, species richness can be increased when there is a larger 'box' (a) or smaller (b) or more flexible 'balls' (c).

The shape of the utilization curve of a species on a resource can also influence packing [26]. Theoretically, leptokurtic (thick-tailed) curves show closer packing relative to niche widths than platykurtic (thin-tailed) curves (Fig. 8.1(e)).

Considerable speculation about the causes of local and latitudinal differences and patterns in species diversity has generated many theories and hypotheses. These all relate, in some way, to the three major components of MacArthur's formula, namely resource heterogeneity (Dr), ecological overlap ($\bar{\alpha}$) and the degree of specialization (Du). Clearly, these are not independent. The following sections review the major hypotheses (Table 8.1), and illustrate the complexity and controversy involved in studying species diversity patterns with two examples.

Table 8.1 Mechanisms for the determination of species richness and their modes of action based on non-equilibrium (N) and equilibrium (E) hypotheses

	Hypothesis or theory	*Mode of action*
(N)	Evolutionary time	Degree of saturation with species
(N)	Ecological time	Degree of saturation with species
(E)	Environmental favourableness	Mean niche width and dimensionality of habitat
(E)	Environmental stability	Mean niche width and resource diversity
(N)	Environmental variability	Degree of allowable niche overlap
(N)	Gradual change	Degree of allowable niche overlap
(E)	Spatial heterogeneity	Mean niche width and resource diversity
(E)	Area	Resource diversity and habitat dimensionality
(E)	Productivity	Mean niche width and resource diversity
(E)	Competition	Mean niche width
(N)	Compensatory mortality	Degree of allowable niche overlap
(E)	Circular network	Degree of allowable niche overlap

8.1 Time
This factor operates on two different scales.

8.1.1 Evolutionary time
Species positions in the community exist, but are only filled if sufficient time is provided for speciation and evolution to occur. Thus species in communities of more recent origin and those that are perennially disrupted, should not have evolved as much interspecies adjustment or speciation as

older, more stable groups of species. As a result they are predicted to be less rich [220]. This line of argument is frequently used to account for the latitudinal patterns of species diversity. The tropical environment is old as it is less influenced by major climatic trends, and this has allowed time for specialization and the evolution of a greater variety of plants and animals (i.e. smaller Du, larger C). Temperate habitats are impoverished with species due to the relatively short time space since recent glaciations and other disturbances. The only unequivocal test of the time hypothesis is to follow the number of species as it changes through evolutionary time. However the fossil record is fragmentary, and such a test is only applicable for a few taxa and certain habitats [11]. For example, Europe is relatively impoverished with tree species compared with North America following the Pleistocene glaciations [28], and at a local level, once-glaciated serpentine mountain areas in Washington have a lower plant species diversity than the unglaciated regions in North California [11]. Diversity also tends to increase during many community successions over time, although in some communities a decrease from late successional to climax stages occurs [137]. Other studies have found no significant differences in species richness between glaciated and unglaciated areas (e.g. trees [6], birds [28]), and as local species saturation does seem to occur (Chapter 7), there appears to be some limit on the time hypothesis.

There is no doubt that worldwide diversity has increased greatly since the beginning of the Paleozoic, as plants and animals have invaded new adaptive zones. As some limit on diversity is apparent after a certain period of time, any further increase in the diversity within an area is likely to be due to an increase in β diversity.

8.1.2 Ecological time

This hypothesis considers more recent time periods (a few to tens of generations), and it is suggested that communities may differ in species diversity due to insufficient time for colonists to disperse into new areas, or for local succession to continue following a small disturbance. Thus newly opened areas (e.g. result of forest fires) may not have the full complement of species. Similarly, isolated or remote areas may be impoverished due to lack of time for colonization. This mechanism may explain the low species diversity of remote islands [61] and isolated mainland habitats (e.g. isolated deserts have a low rodent diversity [58]).

For vagile (widely dispersing) organisms (like birds), historical factors can be largely ignored in continental habitats [11, 61], but in more sedentary organisms, these factors must be considered.

8.2 Environmental conditions
There are several hypotheses which consider that the physical attributes of the environment are the most important in regulating species diversity, either directly or indirectly.

8.2.1 Environmental favourableness
This hypothesis examines the relationship between mean values of environmental variables and diversity (e.g. mean annual rainfall or temperature). It has long been clear that there is some correlation between harshness of climate and reduced species diversity [229]. As a physical factor departs from the optimum for a species, organisms increasingly specialize with regard to this factor, whilst becoming more generalized with respect to other physical or biotic factors (i.e. a larger Du). When all environmental conditions approach optima, organisms can specialize on more gradients, and devote more time and energy to coadaptive adjustments to other species (i.e. smaller Du, large C).

Environmental favourableness has been used to explain the high diversity of tropical communities. The tropics are generally free from protracted storms and have warm and uniform temperatures. Seasonal rainfall does occur, but it has less effect on species than the seasonal temperature changes found in temperate and polar areas [28]. This is supported by several observations, such as more uniform tree growth, less seasonal fruit, and the fact that many birds nest nearly the whole year [224]. It has also been argued that such favourableness may lead to higher speciation rates in the tropics [10,205]. Tropical populations tend to be sedentary, as individuals have little need to move from one locality to another to find a favourable habitat. Topographic boundaries may also pose significant barriers to climatic specialists. Both factors may enhance geographic isolation and reduce gene flow, hence, species—specific characteristics may have evolved more rapidly in the tropics than in less favourable areas where yearly migration and broader tolerance limits are found.

Diversity is generally decreased by chronic environmental stress such as over grazing, air pollution, gamma radiation and harsh climates, where only limited numbers of species are able to cope. Thus, diversity of vascular plants is low in extreme deserts, the high Arctic, salt soils of mangrove swamps and salt marshes etc. [11]. A study of 5902 species of higher plant in California also concluded that environmental favourableness was a major regulator of diversity [228], but on the whole, diversity bears no simple relation to environmental favourableness, as both positive and negative relationships can be found [11].

8.2.2   *Environmental stability*

These hypotheses consider the effect of the variance of environmental variables on species diversity, and bear complex relationships with the three previous factors. The number of species should increase with environmental stability. Stable climates allow evolution of finer specialization (smaller Du's), whereas unstable climates often demand broad tolerance limits favouring organisms with broad niches (large Du's) [9, 28, 228]. Theoretically, competing species tolerate less overlap in a fluctuating environment (i.e. smaller $\overline{\alpha}$ [28]). Stability may also act through the dependability of resources. Fluctuation in resource levels can periodically reduce the size of niche hyperspace (i.e. reduce Dr), subjecting species to periods of intensified competitive squeeze against which the more vulnerable cannot maintain a population (as seen in Chapter 3). Environmental fluctuations increase the probability of a species becoming extinct. Smaller populations which normally have a high probability of chance extinction could then persist longer in relatively constant environments allowing for the accumulation of rare species in the community. Stability in resources could also promote diversity in consumer trophic levels [10], which may, in turn, increase the diversity of the rest of the community (see Chapter 5).

Abyssal depths of the ocean come closer than almost any other environment to having constant conditions. The high diversity of bivalve and polychaete species of these apparently homogeneous areas compared with shallow, more variable coastal regions, led Sanders [230, 231] to propose the Stability—Time hypothesis. This suggests that the underlying cause of high diversity is the persistence of stable environmental conditions over long periods of time, allowing communities to become biologically accommodated with smaller

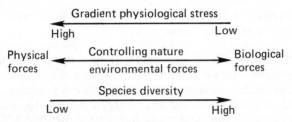

Fig. 8.2 The Stability — Time hypothesis of Sanders [230].

non-overlapping niches. A schematic representation of this hypothesis is shown in Fig. 8.2. The original data used by Sanders have since been re-analysed [232] and serious deficiencies have been revealed. Also alternative explanations have been suggested [1, 232], (see later). MacArthur [229] does provide many examples of inverse relationships between species number of birds, mammals and gastropods and temperature fluctuations during the year (Fig. 8.3), but such relationships are not universal. For example, the diversity of terrestrial vertebrate fauna of Australia and USA show both positive and negative responses to seasonality, depending on the group [15], and the high diversity of coral reefs and tropical rainforests may actually be maintained by disturbances (see below).

Some authorities (e.g. [11, 195]) believe that environmental instability is a major limiting factor on species diversity, acting on the feasibility of marginal niches, competitive effects on niche packing and interactions for niche differentiation. There is evidence for and against this hypothesis but,

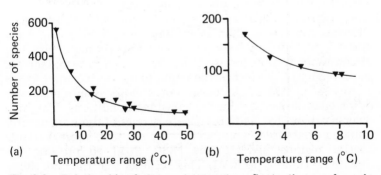

Fig. 8.3 Relationships between temperature fluctuations and species richness in west coast birds (a) and gastropods (b) in America. (After MacArthur [229].)

as described below, environmental instability may facilitate diversity as much as reduce it.

8.2.3   *Environmental variability (or temporal heterogeneity)*
Disturbance rather than stability seems to enhance species richness under certain conditions. Species equilibrium, in accordance with the habitat's carrying capacity, may never be attained if the habitat is frequently disturbed. Such disturbances are indiscriminate, and may remove a large proportion or all the individuals from an area. They include lightning, storms, landslips, flooding, wave action, mowing, trampling, burning and indiscriminate predation and herbivory.

Grime [196] and Connell [137] have proposed that the greatest diversity will occur at moderate or middle ranges on the physical gradient of environmental stress or at intermediate levels of disturbance (the Intermediate Disturbance hypothesis, Fig. 8.4). Under frequent, large disturbances (or severe environmental stress), the community is dominated by opportunistic and rapid colonizers, or species capable of tolerating the form of damage sustained. It is thus of simple structure and low diversity. If disturbances are rare and small, they have little impact and competition leads to the elimination of inferior competitors as the community moves to equilibrium. Here, diversity will depend on resource partitioning or compensatory mechanisms like predation. As the interval between disturbances increases, so does diversity. There is more time for the invasion of species to

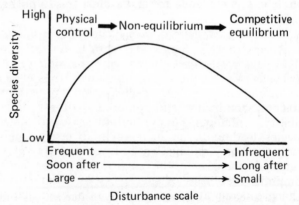

Fig. 8.4   The Intermediate Disturbance hypothesis. (After Connell [137].)

the disturbed area, insufficient time for much competitive elimination to occur, and a greater range of species from all successional stages and of all competitive abilities is maintained (i.e. increased C and $\bar{\alpha}$). Local assemblages are thus kept in a non-equilibrium state (in some sense, supersaturated with species), whereas larger geographical areas might be stable, with minimal species turnover or balanced gains and losses.

The Intermediate Disturbance hypothesis has been suggested as an explanation for the high diversity of tropical rainforest trees and coral reefs [137], herbaceous plants [196] and more recently for arctic alpine fellfield vegetation [233] and cliff bryophyte assemblages [234]. These examples are from trophically simple assemblages of primary space occupiers, but there is evidence for an increase in species diversity of insect parasitoids following habitat disruption [8], so the potential for affecting higher order organisms is there.

Using graphical resource competition models, Tilman [295] predicts the same humped relationship between species richness and distrubance. However, here, the effect is not due to disturbance preventing competitive equilibrium or periodically interrupting competition, but rather by influencing the relative supply ratios of resources for which the species compete. These changes in resource richness and ratios indirectly upset the direction and outcome of competition rather than ameliorating competitive effects directly as suggested by the above authors. Tilman provides substantial support for the model.

8.2.4   Gradual Change hypothesis (climatic predictability)
Many aspects of the climate fluctuate in a regular and predictable way, with a reasonably short periodicity (e.g. day to day, year to year, etc.). Organisms may thus evolve some degree of dependence and specialization on particular environmental conditions and temporal patterns of resource availability, enhancing daily and/or seasonal replacement of species leading to an increase in diversity (increase in function C). The Gradual Change hypothesis [137] suggests that gradual changes in the environment alter the ranking of competitive abilities of species within a community, and the changes occur at a high enough rate, such that the process of competitive exclusion is seldom if ever completed. This again suggests a non-equilibrium community. This is

really an extension of Hutchinson's 'Paradox of the Plankton' ideas [235], developed to explain the diversity of plankton in structurally homogenous waterbodies. Any species can persist, provided that a suitable combination of conditions occurs with sufficient frequency and duration and no one species is able to eliminate others before its own superiority declines. The end product may be a succession of species through the year.

Evidence that this factor enhances species diversity has been found in a variety of taxonomic groups including zooplankton (the *Daphnia curinata* complex [236]), small mammals (cotton rats and prairie voles [60]), fish [244] and northern bumblebee species [237]. Species diversity between distantly related seed predators (insects, birds and small mammals) also parallels the 'Paradox of the Plankton' idea [163]. Even in a chronically variable environment, if it is predictably unpredictable, specialization in irregularity may again be a major means of niche differentiation. This may help to explain high species diversity of perennial and annual plants in the Sonoran Desert [11].

8.2.5 *Spatial or structural heterogeneity*

Habitat segregation is a major means of coexistence between competing species (Chapter 4), so heterogeneous and complex physical environments should contain more complex communities and hence maintain a high species diversity. This increase in heterogeneity should lead to an increase in total niche hypervolume space (Dr), greater habitat specialization (decrease in Du), and generally to an increase in resource partitioning possibilities through added dimensionality to the habitat (increasing C). Theoretically, spatial heterogeneity in the environment can also act as a powerful stabilizing factor tending to oppose destabilizing effects of interspecific competition and predator/prey interactions in simple model systems [238].

At the level of microspatial heterogeneity, involving objects like rocks and vegetation, there is much evidence in support of this hypothesis. Temperate zone deciduous forests offer herbaceous plant species a wider range of light intensities beneath open tree and shrub strata, and thus possess a higher species diversity than evergreen forests which only provide a uniform shade beneath their canopies [140]. In addition, microscale soil surface bumps and cracks [138] and increased soil heterogeneity (laterally and vertically [75, 239]) both enhance plant species diversity. Tilman [295] has been able to model steady state competitive equilibria among 40 species for two limiting resources in a spatially heterogenous environment (high variance in resource availabilities between microhabitants). The complexity of vegetation is in turn an important factor in animal diversity, as found in bird (Fig. 7.6(b)) and insect assemblages [240, 241]. Host plant diversity has also been shown to influence insect species richness among leafhoppers, Lepidoptera, Coleoptera, Diptera and leafminers [218], and to have a major influence on the diversity of African squirrels [104]. Structural heterogeneity of the habitat as a whole has been demonstrated to be an important factor in the regulation of species diversity of numerous groups of animals including waterbugs [132], ants [242], infaunal marine assemblages [243], freshwater fish [244], and many other freshwater taxa [10]. Structural complexity may also enhance species diversity through the moderation of intensive predation (see Section 6.3.1).

On a larger scale, regional diversity depends on the geographical replacement of species populations by others among different habitats, so areas with more habitats are likely to contain more species (high β diversity). For example Columbia is made up of numerous valleys and mountains and is generally geographically complex. As such, it contains more species of most animal types than Brazil and Zaire, which are on the same latitude but topographically simpler [10].

Latitudinal patterns of species richness are often explained on the basis that the tropics are more heterogeneous (e.g. [246]) and hence possess a greater total niche volume (larger *Dr*). For example tropical forests possess a unique variety of vegetative structures such as lianas, vines, and bromeliads, which all provide unique microhabitats. The variety of plants in turn provide a wide variety of flowers and fruits, and in conjunction with reduced seasonality in plant growth and reproduction, this is accompanied by an equally impressive variety of specialist frugivores and seed predators not found in temperate areas. Such producer activity also enhances diversity of monophytophagous insect species. The tropical range of insect sizes is greater than the temperate one, leading to a greater range of bill sizes amongst tropical insectivorous birds [28]. There also appears to be a greater degree of patchiness in the tropics [10, 28], either due to increased competition leading to smaller niches and restricted distributions or possibly through intermediate levels of disturbance (see earlier).

The tropics do seem to show greater heterogeneity than

temperate areas at both local and regional levels, and this is exemplified by a study of land birds by MacArthur [28]. Five acres of Panama forest has 2.5 times as many bird species as five acres of forest in Vermont (USA). An area of 100 000 square miles in Ecuador has seven times as many species as an equal area in New England. This demonstrates that there are both more species per habitat and more habitats per square mile.

Correlations between the structural complexity of habitats and species diversity of their biotas are widespread and usually viewed as the outcome of competition, leading to resource partitioning. There is little doubt that environmental complexity plays an important role in the regulation of species diversity.

8.2.6   The area hypothesis

The importance of area *per se* on species diversity has already been examined in Chapter 7, and many species—area correlations have been documented. Geographically extensive habitats should contain more species than small, peripheral ones, and as the deep sea abyssal regions cover vast areas, the unusually high diversity of marine invertebrates documented by Sanders [230] may simply be an expression of this rather than based on the stability—time hypothesis. Indeed, 99% of the variation in species number found in this study can be accounted for on the basis of a species—area relationship [232].

One explanation of latitudinal diversity gradients is based on area (see [200]). The average area of life zones in which species of grazing ungulates are found, for example, apparently decreases as latitude increases, and this may explain the reduction in the number of species found as one moves from tropical savannah to neotropical regions. This kind of analysis has not been carried out for many other groups, and although area is undoubtedly important at the local level, extrapolation to latitudinal gradients is rather unsound.

8.3   Biotic factors

Several hypotheses stress the role of biological interactions in the production of diversity patterns.

8.3.1   Productivity hypothesis

This suggests that greater productivity should lead to greater species diversity, all else being equal. In habitats with little food, foraging animals should have wider diets [40], but in more productive habitats, greater resource availability can lead to increased dietary specialization without a decrease in population size, so the same spectrum of resources will support more species. Also certain resources, which are too sparse to support a species population in unproductive habitats, may be successfully exploited in productive ones [9]. Thus increased productivity should lead to an increased *Dr* and a reduced *Du*.

Evidence in elusive, although a number of correlations have been noted. Tropical forests are productive and possess more bird species covering a wider range of ecological roles than are found in temperate forests [205]. For example, temperate forests have no obligate fruit eaters (like parrots), birds of prey specializing on reptiles or birds following ant swarms. A direct relationship between species diversity of desert rodents and productivity has also been documented (Fig. 8.5), but this is one of the few empirical demonstrations of such relationships. Increasing productivity, however, does not always lead to increasing diversity. In land plant assemblages, there is no evidence that productivity has a marked independent effect on diversity [11, 228], and when increased productivity is accompanied by a reduction in resource variety as in polluted rivers, species diversity will actually fall. Similar inverse relationships between

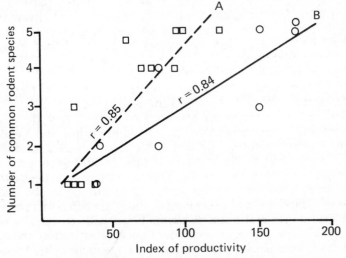

Fig. 8.5 The relationship between rodent number and productivity in the Sonoran desert sandy flatlands (A—○) and Mojave and Great Basin desert sand dunes (B—□). (Adapted from Brown [58].)

primary productivity and species diversity have been found in Californian grasslands and amongst cladoceran consumers in Danish lakes [6]. Theoretical results actually suggest that a humped resource richness—species richness curve may be a general and robust prediction of an equilibrium theory of plant competition for resources, and there is much evidence in support [295]. Thus productivity does not have a simple nor intrinsic relationship with diversity. It is also difficult to isolate from other factors, such as stability in time, and a combination of the stability hypothesis with productivity (stability of primary production) has been suggested [9]. Areas with temporally stable, and/or predictable patterns of productivity may allow component consumer species to partition resources temporally as well as spatially, and thus allow the coexistence of more species than would be possible in areas with more erratic productivity (e.g. in African squirrel guilds [104]). Stability of primary productivity may, in fact, be a more significant determinant of species diversity than productivity *per se*.

8.3.2   Competition hypothesis

This applies to communities which reach equilibrium, and suggests that species composition is a result of past and present interspecies competition. At equilibrium, each species is competitively superior at exploiting a particular subdivision of the habitat or other resource, and diversity is a function of the total range of resources (Dr) and the degree of specialization of species to parts of that range (Du) (the Niche Diversification hypothesis [137]). Without disturbances, species composition will persist at equilibrium and species number may be simply expressed as Dr/Du. After perturbations, the equilibrium is restored.

Two lines of argument are used to suggest that tropical communities are subject to greater competition and hence become more densely packed. The first suggests that as populations will be near maximal size, natural selection for competitive ability (K selection) will be strong and specialists will be at a competitive advantage. Populations in the less-diverse temperate and polar communities are often less stable and below maximal sizes, so competition will frequently be lax. Selection for rapid reproduction (r-selection) is strong and such species often have relatively large niches [9]. Population fluctuations are sometimes as great or greater in the tropics as elsewhere, but this could be explained by large competition coefficients between species [28]. In addition, if evolution has occurred more rapidly in the tropics (see Section 8.2.1) the evolutionary divergence might lead more rapidly to specialization.

The second argument suggests that individuals in temperate and polar zones are selected for their degree of physiological adaptation to extreme conditions [10, 14] hence parameters C and $\bar{\alpha}$ are reduced. Release from rigid physiological constraints allows evolution in the tropics to respond more directly to the biological environment, leading to niche diversification through competition.

There is much circumstantial and some experimental evidence to suggest that tropical species are more highly evolved and possess finer adaptations than temperate ones. For instance, competition between ant species is not sufficiently strong in temperate deserts to lead to the exclusive foraging areas which are maintained by interspecies defence in tropical assemblages [109]. The specialist birds described in Section 8.3.1 and the data in Fig. 3.7 provide further examples. Motile animals living in diverse assemblages do seem to be sufficiently specialized to have achieved a degree of resource partitioning which allows coexistence at equilibrium (Chapter 4). However, there is debate over whether plants and sessile aquatic animals have, or can evolve the required degree of specialization necessary to maintain high diversity in communities like tropical rain forests and coral reefs, given the similarity in general requirements between species [137 vs 295, 298].

As discussed in earlier chapters, it can be shown that competition exists in natural communities and is potentially the most important regulatory factor of species diversity.

8.3.3 Compensatory mortality hypothesis

This predicts that high diversity is maintained when mortality from causes unrelated to competition falls most heavily on whichever species is the most abundant or ranks highest in competitive ability [137]. As seen in Chapter 5, predation may act in this way preventing competitive elimination of inferior species and enhancing species diversity (increasing $\bar{\alpha}$). In contrast to the competition hypothesis, less competition is predicted in more diverse communities than less diverse ones, but complementarity between the two hypotheses is possible (Chapter 6).

Latitudinal trends of species diversity have been explained through the occurrence of more predators and parasites in tropical environments, allowing more prey species which in

turn support more predators [10, 205]. This is not a sufficient hypothesis unless it applies to all trophic levels, and as
seen earlier, the influence of predation is clearly not as simple
as suggested here. In addition, one would expect a greater
prominence of keystone species in the tropics but there is
little or no evidence of this [205]. Predation may be a more
local effect than a global explanation of latitudinal diversity
gradients.

The physical environment can inflict compensatory
mortality on dominant species, as in the case of some corals
[137] and mussels [302], but this is very rare.

8.3.4   Circular network hypothesis

Instead of a normal, linear hierarchy, the competitive hierarchy may be circular (Fig. 8.6(b)). At equilibrium, competitors win some interactions but lose in others, thus
maintaining a higher diversity in the area than otherwise
possible. This has been applied to sessile invertebrates beneath
ledges on coral reefs (see [137]) but has otherwise received
little support. A similar idea involving facilitation has also
been suggested. Species A and B compete moderately, C
competes intensively with B but only slightly with A (Fig.
8.6(c)). Species C may then increase the population growth
rate of species A by suppressing species B. Under such a
situation, diffuse competition experienced by a species
may actually decrease with increased species packing. Such
facilitation has been found amongst desert ants [247],
but otherwise has also received little support.

8.4   Explanations of species diversity patterns in two specific examples

The above relationships between the number of species
and physical or biotic conditions suggest that a number of
important factors influence species diversity and community
structure. These are often difficult to untangle, and their
relative importance varies widely from one community to
another. To illustrate this complexity, the following sections

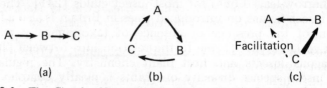

(a)   (b)   (c)

Fig. 8.6 The Circular Network hypothesis. (a) Normal linear competitive hierarchy; (b) circular competitive hierarchy; (c) facilitation.

review the current 'state of the art' in two types of species assemblage.

8.4.1   Insect diversity on plants

Southwood [248] found that the total number of insect species associated with trees was strongly correlated with the cumulative abundance of tree species since the last glaciation, and this work is often quoted as evidence in support of the evolutionary time hypothesis. This interpretation has been disputed by a number of authors [218, 249] who suggest that the variation in species richness is simply a species—area phenomenon (see Chapter 7). The importance of the abundance and/or range of tree species as a determinant of faunal richness has received more recent support [250] and in four different sets of data a significant species—area relationship has been found but no effect of community age on total species number was apparent [220]. However another recent study has again demonstrated a significant correlation between the length of time a tree species has been growing continuously in Britain and the number of insect species [251], but when only native trees are considered, no correlation is found. There is obviously some interaction between host range (area) and evolutionary time (time for species to accumulate [220] and some measure of adaptation to a new host by phytophages [250]), but in the long term, time ceases to contribute and the assemblage appears to be saturated.

The situation is still more complex as some ancient and widespread plant species, such as ferns [240], are significantly impoverished with insect species. The growth form or architecture of the plant may be important, as this can vary the number of microhabitats provided for colonizing insects. There is a hierarchy from tree to shrubs to herbs in both morphological complexity and number of associated insects (see Fig. 7.3). Ferns, which lack complex reproductive structures possess an even poorer insect association [240], and broad-leaved trees maintain 2.3 times as many species as narrow-leaved ones for most insect guilds [241]. The size of insect faunas on introduced trees in Britain is also a function of the presence or absence of taxonomically related native trees [216], due to the relationships between phytophagous insects and host plant chemistry. The regulation of insect species diversity on plants is clearly complex, exemplified by folivorous insects where a host of factors could interact, including autoecological responses to weather, phenology, host chemistry and range, isolation, migration, habitat heterogeneity, the action of predators and parasites and less frequently competition [124].

8.4.2   Rainforest trees

In lowland tropics, up to 100 different species, many rare, can be found in a single hectare [9], and it is difficult to conceive of niche differentiation on primarily non-biotic niche dimensions such as nutrients, vertical height, time and soil depth, that could account for such diversity. The patchy nature of species distributions must also be explained, and to this end numerous hypotheses have been proposed. Of these, the Equal Chance and Circular Network hypotheses [137] have recieved little or no support. The Nutrient Mosaic hypothesis suggests that each tree species gradually depletes it's own peculiar set of nutrient requirements under parent trees, thus preventing seedling establishment, but this has not been tested [9]. There is however, some evidence for the following. The Seed Predation hypothesis [180] predicts that successful recruitment of seeds to seedlings will occur in a ring around, but at some distance from, the parental tree through the action of species-specific seed predators. Other tree species can then establish themselves inside and outside this ring, so that dense stands of one species do not occur. Some evidence in support of this was presented in Chapter 5. A similar mechanism involving

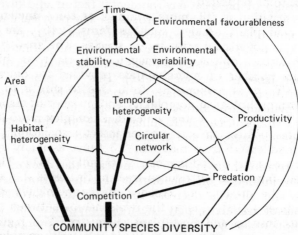

Fig. 8.7 A causal network of factors that may influence the species diversity of a community. Thickened lines indicate the more important interactions.

fungal pathogens has recently been demonstrated [245], but other studies have shown no evidence of compensatory mortality near the parent (see [137]). Intermediate levels of disturbance by fire, floods or storms or epiphyte loads on individual trees, frequently open up patches in the forest, fostering local secondary succession and maintaining a non-equilibrium status of the assemblage which can explain the high diversity. Relatively undisturbed tropical forest areas actually can become dominated by a single tree species [137].

Within a local area, variation in resources and resource ratios may enable many species to coexist through niche differentiation at equilibrium [295] and a certain degree of compensatory mortality probably exists. However for these long-lived tree species (and similarly for corals [137]), the community may be prevented from reaching equilibrium by gradual climatic changes or shorter-term disturbances, which interrupt or alter the competitive interactions, and help maintain the high species diversity.

8.5 Conclusion

In the analysis of species diversity and diversity patterns, one cannot look for a single explanation involving only one causal factor. There is a multitude of ways in which the mechanisms discussed above can interact and have interacted over evolutionary time to produce the assemblages we see today (Fig. 8.7). Conceptually, species diversity may be studied as the relationship of physical environmental factors and diversity, or through the role of biotic processes. The physical environment, including present conditions and past history, determines the pattern of biotic interactions, and ultimately an understanding of diversity patterns requires an autoecological approach (species tolerances, etc.). This will provide knowledge of the action of mechanisms operating through the physical environment. Biotic interactions, in turn, produce patterns in resource partitioning, which are the proximate cause of coexistence and hence observed species diversity [228].

The relationships in Fig. 8.7 indicate that factors which have an influence on the degree of competition between species and allow a decrease in competition through partitioning of resources, play the most important role in the regulation of species diversity and hence in the organization of a community. It is so widely accepted that competition is the major organizing principle in ecology, that it has almost achieved the status of a paradigm.

Chapter 9

The relative abundance of species

Ecologists have long recognized that species are not equally abundant, and that communities with similar species densities often differ in another way; the relative importance or abundance of the constituent species. Considerable effort has been expended in attempts to document differences and regularities in relative abundances, as an understanding of these may well provide an insight into the structure of natural communities [210].

The relative importance of a species can be examined in a number of ways. For animals, population density or species biomass is used, whereas for plants, coverage, frequency or basal area are also important [11]. Whichever method is used, some species are found to be rare (low importance), a few common (most important) and many species are of varying intermediate degrees of rareness, even under conditions of considerable environmental uniformity [6, 209]. Various methods have been used to graphically display these relations, and several models have been suggested to account for the distribution of species importance found within communities. The characteristic features of the different models have been reviewed by Whittaker [11], so the following sections will only examine these briefly. A more mathematical approach can be found elsewhere [209, 252].

9.1 Fisher's series (log series)

Plots of the number of species against the number of individuals per species often provide a characteristic distribution

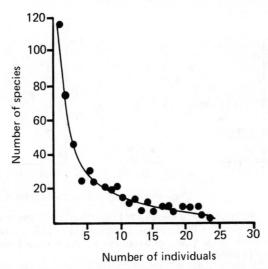

Fig. 9.1 Distribution frequencies of Malayan butterflies. (After Fisher *et al.* [253].)

pattern. This was first mathematically described by Fisher *et al.* [253] from light-trap studies of Lepidoptera (Fig. 9.1). This pattern followed a logarithmic series,

$$S = n_1\left(1 + \frac{x}{2} + \frac{x^2}{3} + \ldots\right)$$

where n_1 is the number of species represented by a single specimen, $n_1 x/2$ by two specimens etc., and x is a constant that approaches unity in large samples. This model provides a reasonable fit to some types of samples and it fits the ascending slope of high importance values of the lognormal distribution (see later). However, no theoretical justification is assumed for the series and so this model has only rarely been applied [e.g. 317].

9.2 Lognormal distribution

Preston [254] found that species abundance data from large collections of animals fitted a lognormal distribution when the number of species was plotted against the number of individuals combined into octaves, each succeeding octave containing a range twice that of the previous one. (Species that fall on the boundary of octaves are divided equally between the two.) This suggests that species are distributed

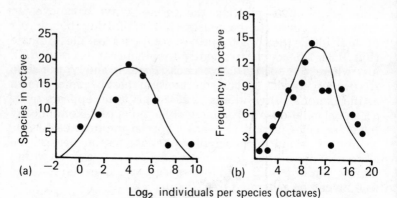

Fig. 9.2   Examples of the lognormal distribution of species abundance: (a) 10-year breeding bird census. (After Preston [255]); (b) combined net productivity of plant populations of three mixed forests. (After Whittaker [140].)

randomly about some mean value on a $\log_2$ scale of abundance (Fig. 9.2). The distribution is given by

$$N_R = N_O\, e^{-(\frac{1}{2})(R\sigma)^2}$$

where N_R is the number of species in an octave, R octaves distant from the modal octave (N_O), and σ is the standard deviation of the distribution.

The pattern has been documented for many disparate groups of organisms, provided that the sample is large enough to enclose the modal abundance class (normally > 100 species). These include birds [255], diatoms [206], bees and flowers [256], and ants [242], and the pattern is now recognized as a general rule for collections of large numbers of taxonomically related species [210]. The dispersion of the curves (σ) is fairly similar for various groups of organisms, e.g. 2—3 for birds, 3.1—4.7 for moths [254]. The values also vary with the environment. For example, for forest birds, σ = 0.98 in lowland tropics, 1.36 in temperate regions and 1.97 for islands [224]. Similarly, for diatom assemblages in different mainland and island streams, σ ranges between 2.49 and 5.1 [206]. Both studies suggests that there are fewer moderately common species and more abundant species on islands than on mainland areas, supporting the ideas explored in Chapter 3.

In addition to plotting the species curve as above, an individual's curve can be constructed by plotting the number of individuals that each species comprises as the ordinate (Fig. 9.3). This curve frequently terminates at its crest (RN), just where the species curve ends (Rmax), and γ, the ratio RN/Rmax equals 1. This relationship has become known as the Canonical hypothesis [225]. This again appears to be a general relationship for large collections of taxonomically related species, and it also has a special importance for the theory of island biogeography. As a consequence of the Canonical hypothesis, the function $a\,((2\sigma^2)^{-\frac{1}{2}})$ is fixed within narrow limits (≈ 0.2) and the species area constant Z equal to 0.25 can be generated (see Chapter 7).

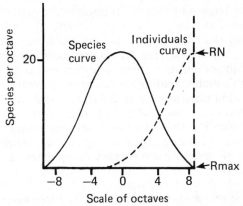

Fig. 9.3 The canonical lognormal distribution for an assemblage of 178 bird species, adjusted to have a mean of zero. (After Preston [255].)

However, there is much debate over the interpretation of these relationships in functional terms. They are not applicable to all situations, especially small samples with few species. As the number of species increases, so the number of factors governing their relative importances will increase. From a statistical basis, the Central Limit theorem predicts a lognormal distribution for a large collection of species affected by diverse, independent factors that are compounded multiplicatively [209, 256]. It has thus been suggested that the lognormal distribution reflects statistical laws of large numbers and has nothing to do with the underlying biology (e.g. [85, 209]). It has also been suggested that the observed values of the two canonical parameters a and γ

are simply a robust consequence of lognormal distributions in general [209]. Although this is probably true of the rule $a = 0.2$, Sugihara [210] has shown that the canonical hypothesis $\gamma = 1$ is more closely conformed to by biological communities than is explained by mathematical generalities and so reflects real regularities in the shape of the lognormal distribution in these communities.

A number of recent models have been suggested to explain the lognormal distribution of species abundance. The first assumes that the underlying structure of niches is reflected in the relative abundance pattern, and that minimal niche structure for communities is hierarchical so that one can sort species into natural groups according to increasing niche similarity [210]. If total communal niche space is sequentially split up by large numbers of taxonomically related component species, so that each fragment denotes the relative species abundance, then the sequential fractures can lead asymptotically to a lognormal size frequency distribution. Speciation may be considered as a successive carving up and elaboration of a taxon's niche, leading to community structure consisting of subdivided taxonomic guilds. This model is similar in spirit to MacArthur's (see later) but differs in two important respects. Firstly, the breakages are sequential rather than simultaneous, leading to the lognormal distribution, and secondly this model incorporates complex random breakages

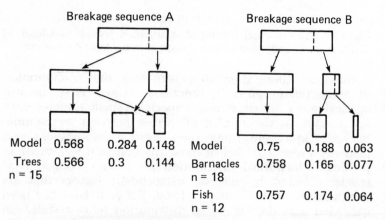

	Breakage sequence A				Breakage sequence B		
Model	0.568	0.284	0.148	Model	0.75	0.188	0.063
Trees n = 15	0.566	0.3	0.144	Barnacles n = 18	0.758	0.165	0.077
				Fish n = 12	0.757	0.174	0.064

Fig. 9.4 Two possible breakage sequences in Sugihara's Minimal Community Structure model with associated data from real species assemblages. (After Sugihara [210].)

involving the translation of several niche axes to species abundance. Some limited evidence is presented in support of the model (Fig. 9.4). Sugihara concludes that this model not only explains the two ubiquitous patterns of the canonical lognormal and the species area constant, but also is intuitively consistent with evolutionary conditions for generating species diversity. Pielou [252] describes a 'resource opportunity' model based on sequential and random breakages of a single resource axis, and this again leads to a lognormal distribution.

These models do not allow species to change octaves in a dynamic way, and they assume that all species have the same probability densities of abundance, i.e. that they are symmetric. As an alternative, Ugland and Gray have described an asymmetric model [257]. Most communities consist of patches, and the frequency distribution of species within any patch is unpredictable, but stabilizes over the whole community when the patches are summated. The model assumes that communities consist of three abundance groups (rare (65%), intermediate (25%), common (10%)), and within each group the species abundance relation is symmetric. An asymmetric model is built up from the sum of such symmetric models, providing a good fit to the lognormal. Thus the reason why the lognormal appears in the model is purely statistical. Such a fit describes an equilibrium state within the community (one of many possible states) which is dynamic. For example, marine benthic community data over four years showed good fits to the lognormal, in spite of different species dominating each year [257]. This model will only apply to large samples where the abundance distributions are summed over many patches, but it lacks any fundamental biological explanation for the distribution.

The lognormal distribution is very successful at describing species abundance patterns in large collections of species, but at the moment, it is hard to ascribe to it any sound biological interpretation. It appears that as the heterogeneity of a sample increases under (a) increasing sample randomness, (b) the influence of more environmental factors, and (c) a decreasing influence of species interactions, then the relative abundance distribution approaches a lognormal pattern.

9.3 'Broken stick' or random niche boundary hypothesis

MacArthur [258] developed this model based on the biological assumption that the abundance of each species is determined by random partitioning of an important niche parameter with no overlap between species. All resources are assumed to be utilized, the competitive ability of each species is a random variable and the number of species is a premise of the model, rather than a prediction. Hypothetically, to predict the relative abundance of S species, the resources can be considered to be distributed along the length of a stick, on which $S - 1$ points are picked at random and the stick broken at these points. The length of the resulting segments is proportional to the relative abundance (or niche size) of each species, and can be given by the series

$$Pr = \frac{N}{S} \sum_{i=1}^{r} 1/(S - i + 1)$$

where S is the number of species in N individuals per sample; i is the sequence of species from least to most important and r is the position of the species with a relative abundance of Pr in the sequence. When the relative abundances are arranged on a log scale in decreasing rank order, the distribution is nearly always linear (Fig. 9.5). The model, as designed, corresponds to a community in which species compete, each one excluding all others from the resources it exploits.

Since the publication of the original model, data have accumulated which both confirm and reject its validity [259, 260]. The best fits seems to be found among species in fairly stable equilibria, characterized by large body size and long life cycles [259] and of high taxonomic affinity in competitive contact with each other [11]. Birds, predatory

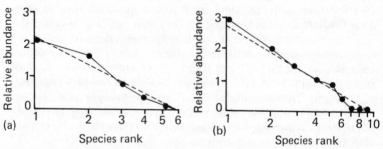

Fig. 9.5 Examples of the Broken Stick distribution of species abundance. (a) Assemblage of fish (family Percidae); (b) ophiuroids (brittle stars). Dashed line is the expected relationship, solid line the observed. (After King [259].)

gastropods and fish often have relative abundances which fit the broken-stick model. Small-bodied, relatively short-lived or opportunistic species with fluctuating populations generally do not [e.g. 318]. The broken-stick model has never been successfully used for plant assemblages [11].

The accuracy of this model has been questioned, as by and large, it tends to underestimate the most abundant species and overestimate the least abundant [261]. Mac-Arthur also indicated that he no longer believed his model to be realistic [262]. The same distribution of relative abundances can be predicted from models not including competition, or any species interactions, among their assumptions. An 'exponential' model of Cohen [263] and an 'open community' model of Ricklefs [10] both conform to the 'broken stick' distribution. If the underlying picture of species abundances is the result of an intrinsically even division of some major environmental resource, then the statistical outcome is the 'broken stick' distribution [209]. Data fitting the model do not validate the specific model of MacArthur, but do indicate that some major resource is being roughly evenly divided amongst the constituent species of the community. This contrasts with the interplay of many independent factors inherent in the lognormal distribution of species abundance.

9.4 Niche pre-emption or Geometric Series hypothesis

In contrast to the even, random distribution of resources as suggested by the 'broken stick' distribution, Whittaker [140] introduced the idea of 'niche pre-emption'. Again, the community is dominated by a single factor, but the sizes of niche hypervolumes (and hence species relative abundances) are determined by certain species pre-empting part of the niche space, leaving the remaining species to occupy the rest. The first or dominant species in the sequence occupies a fraction K of resource hyperspace, the second species a fraction K of hyperspace not occupied by the first, and so on. Niche pre-emption has an ensuing species importance distribution following a geometric series [209], where the importance of species i (P_i) in the sequence from most to least important is given by

$$P_i = P_i C^{i-1}$$

where C is the geometric species ratio (relative abundance of species i/relative abundance of predecessor). This approximates to $1-K$ [11]. The series continues until the rarest species contributes to all the remaining abundance. Under certain circumstances, niche pre-emption can give a log series distribution [209]. There is no assumption that the dominant species is the first colonizer, but the model assumes that the species are in some form of competitive balance, and take roughly equal fractions of resources not claimed by more effective competitors. This implies strong dominance in the community, so rare species will be much rarer for a given sample size than in the lognormal case. If log species importance is plotted against species rank, a linear relationship is produced by this model (Fig. 9.6). This model does have a sound biological basis. It can successfully be applied to vascular plant assemblages in early successional stages or in harsh environments, where few species and strong dominance are found, producing a steep curve [140]. A less steep curve is found for some assemblages of moderate diversity in less severe environments. Fractions of many communities of higher diversity (e.g. each plant stratum in a forest) also fit this distribution [6, 11] and rank abundance curves of insect groups follow a geometric series distribution in early stages of succession [264].

9.5 Conclusion
These different models form a range of intergrading curves which can clearly be seen by simultaneously plotting the predicted distributions of log species abundance against

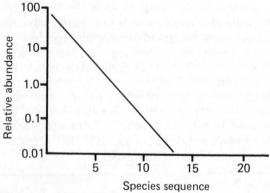

Fig. 9.6 The Geometric series distribution of relative abundance. (After Whittaker [140].)

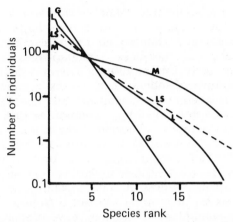

Fig. 9.7 A simultaneous plot of species abundance distributions as predicted by the four different models. G — Geometric series; L — Lognormal; LS — Logseries; M — Broken-Stick. (After Whittaker [11].)

species rank (Fig. 9.7). Where the dynamics of the community are dominated by some single factor, the 'broken stick' model is appropriate as a statistically realistic expression of a uniform distribution and the log series is a statistical expression of an uneven niche pre-emption process of which an ideal form is described by the geometric series [209]. Intermediate between these extremes is the lognormal distribution, produced in a randomly fluctuating environment and/or when several factors are important in the dynamics of a large assemblage of species. A fit to one or other of these models must express something significant about the groups of species that the distribution represents. Instead of choosing among the models for the best description of species relative abundances, one must enquire when and why the relative abundances of species in a community approach one of these models.

The 'broken stick' distribution is to be expected whenever a small, ecologically homogeneous group of species divides a fixed amount of some major resource randomly amongst themselves. Competition is not necessarily the regulating factor, but a fit to this model indicates that a single factor predominates. The resulting species abundance distribution is significantly more even than in the lognormal.

If the community is again dominated by some single factor, but divides niche volume in a hierarchical fashion, a log series or geometric distribution is expected, which produces a significantly less even abundance distribution than the lognormal. Simple communities showing strong dominance as found in early successional stages, harsh climates or sections of larger communities, are described in this way.

Once a number of independent factors become important, the pattern of relative abundance takes on the lognormal form. This is approached by communities that are rich in species, or by samples of species from a range of environments and communities. The biological basis is rather obscure, and the lognormal may not provide information about ecological interactions among populations in the same community, as many species in such a sample may never co-occur. Indeed the lognormal is formed from a combination of smaller samples which can have geometric, lognormal or 'broken stick' abundance distribution [11].

The geometric and lognormal distributions may actually represent extremes in sample size. For example, when all biomass data from Californian grasslands are pooled, a lognormal distribution is found. As individual stands are progressively separated, a progressively poorer fit to the lognormal and closer approximation to the geometric series results [6]. Similarly, two adjacent grab samples from a marine benthic community usually show quite different dominance patterns, but when five or more samples are pooled, a lognormal distribution results [257]. Small-scale patterns are influenced more by biological factors such as predation, competition, recruitment, differential mortality etc., and so the geometric distribution may describe partitioning of realized niche space among co-occuring populations or within guilds. The lognormal probably describes partitioning of realized niche space amongst species guilds, or even communities.

A change in species abundance distributions is also apparent during succession. Early successional stages fit the geometric series, but as succession proceeds, a multitude of ecological dimensions are likely to be relevant to the ultimate composition of the community, resulting in a shift towards a lognormal species abundance distribution (Fig. 9.8). This has been demonstrated for plant succession (although late successional stages may revert back to dominance and a geometric series [11]), and in insect groups whose species abundance distributions gradually became more even as succession proceeds [264]. This shift may be a direct consequence of increasing habitat heterogeneity along the lines

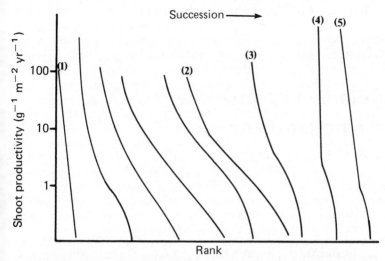

Fig. 9.8 Species importance distributions for an old field succession to oak—pine forest. The curve of the initial old field community (1) is geometric. At the later herbaceous stage (2), the curve approaches a lognormal distribution. Woody plants enter the assemblage at (3) and gradually establish strong dominance (4) leading to a reduction in species diversity. The beginning of a secondary increase in diversity is seen at (5). (After Whittaker [11].)

of the community equilibrium model discussed earlier [257]. A reversal of this pattern is seen when mature communities become polluted [7], and adaptation to a single dominating factor is again necessary.

The fact that a lognormal distribution is produced by combining geometric series suggest that few species combine (a) wide occurrence with high abundances, or (b) limited occurrence with rarity, and that most species, (c) occur with medium abundance in a medium number of communities, (d) have high abundance in few communities, or (e) have low abundances in a large number of communities [6].

The study of relative abundance distributions has led to an increasing recognition of the limitations of their curves for interpreting community relationships, and has given less insight into community organization than had been hoped. Nevertheless, these studies have shed some light on the relationships of community structure and function consistent with other approaches (e.g. Chapters 3 and 6), and new models on the lines of Sugihara's seem to hold some promise for the future.

Chapter 10

Community structure: the patterns and rules

Community structure embodies all the various ways in which individual members of communities relate to, and interact with, one another to produce patterns of resource allocation and spatial and temporal abundance among constituent species. Using two indices of community organization, namely species richness and relative abundance, I have tried to answer a number of questions which may help to identify the patterns and rules associated with the structure of natural communities (Chapter 1).

10.1 Community patterns

Have any patterns in community organization actually emerged? I believe that the answer is yes, and the most obvious of these are described below.

(a) Interspecific interactions (namely competition and predation) generally constrain the niche of each species below limits imposed by physiology or morphology.
(b) Intraspecific competition tends to increase niche width, interspecific competition tends to decrease it.
(c) When species are very similar, or competition intensifies with decreasing resources, competitive exclusion of species from areas of overlap may occur.
(d) Ecologically significant size differences can exist within and between species in a guild when they are separating on one major resource gradient.
(e) Density compensation or ecological release occurs in depauperate guilds.
(f) Niche shift or character displacement can occur when

species similar in allopatry overlap in range in sympatry.
(g) Coexisting species show niche separation; animals utilizing three major dimensions (temporal, spatial and trophic) and plants generally utilizing two (temporal and spatial).
(h) Predation has a greater impact on plant than on animal assemblages but the effect on species diversity depends on the number of strongly interacting levels.
(i) Communities become saturated with species over ecological or evolutionary time.
(j) At the gross geographic level, species replace one another horizontally from one habitat to another.
(k) Most groups of organisms show latitudinal trends in species richness.
(l) Large areas support more species than small areas.
(m) Complex habitats support more species than simple ones.
(n) Persistent habitats tends to support more species, with narrower niches than ephemeral habitats.
(o) Species abundances from large collections of taxonomically related species tend to follow a canonical log-normal distribution.
(p) In small collections of related species or in simple communities, one dominating factor or resource tends to control species relative abundances. This leads to a 'broken stick' or geometric distribution depending on whether even or uneven division of resources occurs.
(q) Few species have a wide occurence and high abundance or limited occurrence and are rare. Most species have a medium abundance in a medium number of communities, high abundance in a few or low abundance in many communities.

The competition hypothesis provides the most plausable explanation for most of these patterns. This hypothesis does, however, assume that communities are structured by interactions between species.

10.2 Communities: random or structured species associations

Are communities actually structured or simply random collections of non-interacting organisms? Any random assemblage of many non-identical objects has some properties that give it the appearance of structure even if the objects are not interacting. This is due to the statistics of large numbers.

However, there are many examples where the world appears chaotic at the level of the individual species but constant and predictable at the level of community organization [7]. There is also evidence to suggest that communities are not randomly assembled, and mathematical methods which can statistically prove that they are not.

10.2.1 Overdispersion of niches

If limits on niche similarity of coexisting species are coupled with the effects of interspecific competition between them, then a regular (equally spaced) separation of species in niche space should be apparent (in effect a balancing of two forces). Where this is so, the alternative or 'null' hypothesis of randomly generated differences between niches must be rejected. This has been termed overdispersion of niches [136]. On a single resource dimension, Hutchinson's 'rule' provides evidence that species are ordered in a non-random fashion along the resource. A more detailed study over three niche dimensions among ten grassland bird assemblages has demonstrated that although the relative importance of various dimensions in separating the species differs in various assemblages, overall niche separation is remarkably constant

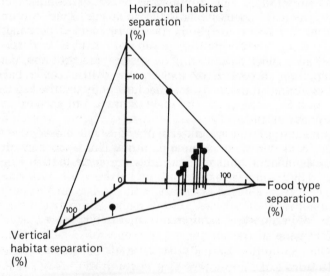

Fig. 10.1 Average niche separation along three niche dimensions in ten grassland bird assemblages. Squares are South American assemblages, circles are North American. (After Cody [135].)

(Fig. 10.1). This suggests that the birds have reached some sort of limiting similarity and that their niches are over-dispersed in total niche space. Patterns of overdispersion have also been found amongst assemblages of lizards [41, 107, 136], bats [66] and intestinal parasites [318].

10.2.2 Species distributions on islands

Surveys of resident bird distributions on dozens of islands in many archipelagos normally show a significantly non-random distribution of islands per species when compared with randomly generated associations [85, 213]. Also studies on the turnover frequency of species on islands show that more species turnover rapidly or exhibit no turn-over than would be the case if a species differed in turnover rate only randomly [85]. The non-randomness of these patterns can partly be explained in terms of species' habitat requirements, area requirements (minimum population or territory size) and differences in immigration and extinction rates between species. However, these factors cannot explain the innumerable instances where species are absent from suitable islands, so species interactions must be important. This conclusion has also been reached through the develop-ment of a shared island null hypothesis, which allows com-putation of the probability that two species share a number of islands by chance [266] (assuming perfect dispersion). Analysis of the distribution data from several bird species pairs and more complex five species assemblages within archipelagos shows that species are distributed regularly and competitive interactions may have affected the coloniza-tion process. Similar studies of lizard and bird species on 34 groups of islands have also shown a highly non-random distribution of these species [214]. Further evidence has come from surveys and experimental introductions involving ant assemblages on small mangrove islands in the Florida Keys [292]. Species distribution patterns are not due to chance effects and the assemblages themselves are not random combinations of species. Interference competition in the form of aggressive interactions proscribes coexistence of certain pairs of species and the presence of so-called primary species is found to be a critical factor preventing invasion of an otherwise suitable island by so-called secondary species.

10.2.3 Assembly rules

If communities are randomly assembled, then all possible

combinations of species from a species pool are possible on islands within an archipelago or within locally sympatric groups in small areas of habitat. In fact, a number of studies have found that only certain combinations of species actually occur, and this again emphasizes the non-random nature of many species assemblages. Community assembly of island bird species seems to involve the following patterns [213].

(a) Only certain combinations of related species coexist in nature.
(b) Permissible combinations resist invaders that would transform them into forbidden combinations.
(c) Combinations stable on large or species rich islands may be unstable on smaller or species poor ones.
(d) Combinations may resist an invader on small, species-poor islands that would be incorporated on to large, species-rich ones.
(e) Some pairs of species never coexist.
(f) Some pairs of species form unstable combinations by themselves but may form part of a stable larger combination.
(g) Conversely, some combinations composed entirely of stable subcombinations may be unstable themselves.

A study of desert heteromyid rodent species predicted that there were eleven possible species combinations, but only four were ever observed [108]. The conclusion is that the rodents assembled into the locally sympatric groups are not a random subset of the fauna available.

The occurrence of species combinations is constrained by the resources available, the niche characteristics of each species and the pattern of niche separation. The limited resource space in a patch of habitat may be best utilized by a specific combination of species which may leave fewer resources unutilized than a forbidden combination of the same number of species [108, 213]. Thus after a certain period of evolutionary time, the observed assemblies are those showing the maximum possible packing of niches into resource space delimited by the resources available and limiting similarity. These saturated assemblies can not be invaded by other species, whereas, imaginary assemblies based on random associations are undersaturated [108].

10.2.4   Community convergence

If species are assembled non-randomly into communities,

then the patterns of structure of communities from different geographical areas that contain independently evolved biotas but that provide similar physical conditions, should show convergence [7, 213]. For example, the phenomenon of the biome indicates similar responses to environmental factors amongst communities. Rain forests in Africa and South America are inhabited by plants and animals of different evolutionary origins, but of similar adaptations. The wide occurrence of latitudinal species diversity trends also indicates some convergence in the structuring of communities.

On a smaller scale, there is a lot of evidence to suggest that convergence of bird assemblages from different continents does occur (e.g. Figs 7.6, and 10.1). This phenomenon can ultimately lead to the convergence of the physical appearance of members of the assemblages, which is derived from the underlying similarity in the way that resources are divided up (Fig. 10.2). Coral reef fish assemblages show a high degree of similarity at different sites in large natural patch reefs in the Atlantic and Pacific Oceans [271] and similarity over space and time at sites of similar habitat type in the Australian Great Barrier Reef [134]. Granivorous rodent assemblages from readily colonized sandy habitats in different deserts also demonstrate remarkable parallels in the size and foraging techniques of their species even though they may belong to different genera and families (Fig. 10.3). The smallest species are almost identical in size, but to allow the inclusion of an additional species of small size, other members of the Sonoran guild are displaced to larger sizes than their ecological counterparts in the Great Basin. This type of small difference in species packing levels between comparable assemblages has

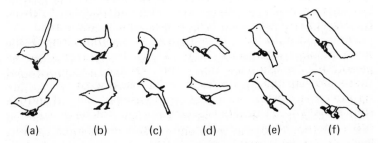

Fig. 10.2 Convergence in physical appearance and size of four canopy insectivores ((a)−(d)) and two sallying flycatchers ((e) and (f)) in the Californian Chapparral (top) and Chilean Matorral (bottom). (After Cody [270].)

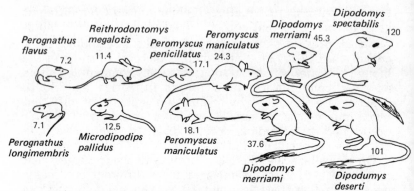

Fig. 10.3 Convergence in structure between a six species rodent assemblage from the Sonoran desert (top) and a five species assemblage from the Great Basin desert (bottom). Numbers indicate average weight in grams. (After Brown [58].)

also been found among birds. For example, African scrub supports between one and five more species than equivalent habitats in California and Chile [42]. However, the total avian or granivorous rodent niche space in different geographical regions may be very similar. There is a limited number of ways in which this space can be exploited, and each species has its own way of doing things, limited in part by what the other species in the assemblage are doing. Chance, historical accidents caused by speciation and geographic barriers, differences in the ages of regions or the configuration of surrounding habitats will affect which species are found in different communities and may cause species impoverishment. Existing species may then utilize at least part of the resources normally used by the missing species and this may explain the small differences in structure frequently found between comparable communities.

However, there are examples where the total niche space for a certain type of organism or assemblage differs in the different areas being compared. This is seen in the profound ecological differences between the terrestrial vertebrate faunas of the USA and those of Australia [14]. Australian lizards occupy a broad range of niches, having usurped many of the ecological roles undertaken by mammals, birds, snakes and predatory arthropods in North America. For example, the average number of desert lizard species in North America is 8, in the Kalahari desert of South Africa 15, and in Australia 28 [107]. This variation in species number is accommodated by a change in the variety of resources utilized by the entire assemblage rather than through a decrease in the average niche size or degree of overlap between the niches of individual species (Fig. 10.4). Indeed, the average morphological nearest neighbour distance is greatest in Australia, so the morphological structure is more open in Australian assemblages than on the other two continents [10]. One might speculate that the assemblages or guilds themselves are subject to competition in resource space with other types of organism or guild. Where there are fewer other types (for historical or other reasons) as in Australia (fewer snakes, mammals and birds), a particular guild or assemblage may undergo competitive release and possible adaptive radiation by expanding the total niche space used. In other localities, where there are many different types of organism (as in North America), particular guilds or assemblages would be 'boxed in' in niche space as niche packing and diffuse competition are high. Any increase in species diversity can only occur through a reduction in niche breadth and/or an increase in niche overlap. Comparisons of communities in these different localities will obviously not provide evidence of convergence, even though they are both fundamentally organized by the same factor, namely competition.

10.2.5 How are species added to communities?

At the beginning of Chapter 8, it was suggested that communities can differ in species diversity in three, potentially interacting ways; the diversity of available resources, niche width of component species and the degree of niche overlap. However, the relative importance of these factors seems to change with the degree of saturation of the communities being compared. This is exemplified by a study of bird diveristy among the Pacific islands [57]. Small islands hold between 30 and 50 species, with a weight ratio of nearest neighbour pairs of approximately 4. These birds have wide niches so require large interspecies differences to reduce competition. On larger islands, bird diversity increases to around 100 species, the extra species being accommodated by an increase in niche overlap, as weight ratios of nearest neighbour pairs have been compressed to 2. Species are more closely packed, but reduce competition by reducing niche width to some degree. On New Guinea, there are 513 species, but the average weight ratio between neighbours has been compressed no further, so no further niche overlap has occurred. Extra species have been accommodated by an expansion of the size sequences and hence of the diversity of resources utilized by the assemblage (more smaller and larger birds), and also by a finer subdivision of space or foraging technique (smaller niches).

Increases in species diversity simply through increasing niche overlap of species is not often reported. As discussed in Chapter 3, the Niche Overlap hypothesis predicts that overlap should actually decrease as species diversity increases. The addition of species to a community is therefore accomplished by a reduction in niche width and/or an expansion of the resource base. There are many examples showing a decrease in niche width and overlap with increasing species diversity but little change in the resource base. These include trees, grasses and shrubs (Fig. 3.7), plant species in old field successions [75], tree invertebrate fauna [250], *Anolis* lizard guilds [26] and birds [42]. Other examples provide evidence of an increase in the resource base with little change in niche width as diversity increases. This is seen in the lizard faunas discussed earlier, where the total volume of lizard niche space tends to increase in areas of more diverse saurofaunas (Fig. 10.4), so that species diversity is roughly proportional to the diversity of resources utilized. Analysis of morphological characters of bird [10, 43, 319] and bat [66, 118] assemblages, indicates that nearest neighbour distances are relatively fixed despite differences in species diversity, whereas their total morphological niche volume increases in direct relation to the number of species. Analysis of appearance and behaviour characters in moths comes to the same conclusion [10]. These results suggest that there is a limiting degree of similarity beyond which competition precludes coexistence of species. Species are then added to assemblages or guilds only when ecological opportunities are increased and the assemblages can expand their total niche space and variety of ecological roles. In conclusion, it appears that an increase in diversity can be accomplished by an increased overlap between species only when assemblages are markedly unsaturated, up to a maximal tolerable niche overlap. Any further increase in species number requires increased species packing through decreases in niche width. This will also have some limit when the guild or assemblage becomes saturated and optimum use is being made of available resources. Further expansion of species

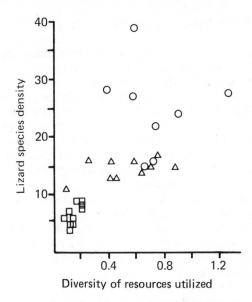

Fig. 10.4 The relationship between lizard species density and overall diversity of resources exploited along three niche dimensions (food, microhabit and time of activity). Squares — North America; triangles — Kalahari desert, Africa; circles — Australia. (After Pianka [41].)

richness of the guild or assemblage can only occur through an expansion of its resource base.

10.2.6 Neutral models

These models of community structure have provided a null hypothesis against which the patterns found in community analysis can be tested. Some are constructed by randomly choosing a set of potentially competing species, or by randomly placing a set of utilization curves on a resource axis (e.g. [267, 320]). If competition has caused the evolutionary divergence of species, then species similarity in an observed assemblage is expected to be significantly lower than that in the neutral, random assemblage. Some analyses have found that species overlap or separation does not differ markedly from that in such randomized analogues (e.g. grasshopper assemblages [268]). One must conclude that competition has not been a force in organizing these assemblages. On the other hand, many studies have shown that mean similarity and average overlap between species is substantially lower in the observed assemblages, suggesting that competition is important in their organization (e.g. lizard assemblages [267, 321]).

Another approach to the neutral model is to study the likely distribution of relative abundance of individuals among S species, assuming no biological interactions [272]. Real assemblages of birds, fish, insects and plants show fewer species or greater dominance of a few common species than is predicted by the neutral model. This discrepancy is greatest among tropical assemblages, where biotic effects are perhaps more pronounced.

This approach has now been exploited in a number of studies of community structure, but there is a hot debate over the usefulness and validity of such null models, especially those assuming non-interactive random communities as the only logical alternative to interactive structured ones [e.g. 320, 322, vs 301, 323].

10.3 Community rules

A number of 'ground rules' for community organization are apparent, which lead, in turn, to the patterns discussed in earlier sections.

(a) Communities do not have an ever-increasing species diversity. In equilibrium communities, the maximum number of species is set by the total niche space available and a limiting similarity between species set by competition. The degree of limiting similarity can be identified only when species differentiate on a single resource axis. For characters associated with the food resource, animal species must differ by a factor of 1.3 in size or 2 in weight. If species separation in niche space occurs on more than one dimension, limiting similarity still exists, but it is difficult to quantify. Niche space can be subdivided within the limits set by a maximal or optimal level of utilization of available resources by species. In non-equilibrium communities, species saturation and limiting similarity are difficult to identify. Any limit to species number will be based on the relationship between resource supply and demand, and whether the non-equilibrium state is a permanent or transitory feature of the community.

(b) An increase in the variety of resources available for specialization occurs under conditions of relaxed interspecific competition (e.g. impoverished biotas). Under the Niche Variation hypothesis, an increase in phenotypic variation (and niche width) within a species population will occur, provided that the species niche has a high between-phenotype component.

(c) As a general rule, increased separation of niches between species pairs occurs as species diversity increases (the Niche Overlap hypothesis). This is accompanied by a decrease in niche width and species population size, up to some limit imposed by minimum resource requirements and/or population size. However, diffuse competition and total overlap between species does increase with diversity. The increased separation of species pairs with increased diversity many represent an equilibration of the overall level of competition that any species suffers, again implying a limiting similarity between species.

(d) Community saturation is further identified through the species—area relationship, and the constancy of the parameter Z (the slope of the relationship) under different situations suggests a further rule in community organization.

(e) The number of niches, niche width and niche overlap are very important in the consideration of the number of species in a community. Any factors influencing these parameters are therefore important factors in community organization (see Chapter 8). Heterogeneity provides much of the explanation of species richness simply through its effects on the number of potential niches and how finely niche space may be sub-divided. Competition is probably the most important factor in motile animals. For sessile animals and plants, other factors may be equally important in the control of species richness, because it is difficult for such organisms to evolve the required degree of specialization to avoid competition. Where diversity amongst these organisms is high, factors such as environmental variability and compensatory mortality are probably important.

10.4 Expressions of community structure

There are a number of characteristics of communities which may be interpreted as providing some glimpse of the nature of their structure.

10.4.1 *The length of food chains*

The number of trophic levels in a food chain is typically

three or four, which is in sharp contrast to the great variability in the amount of energy flowing through different ecological systems. Food chains are no shorter in the barren polar environments than in the productive tropical coral reefs. A number of hypotheses have been proposed to explain this constancy, including those based on energy flow, optional foraging and systems dynamics. Details can be found in a number of reviews [273—275]. Food chains must also have at least one producer in the system, and no three-species food chain loops occur [9].

10.4.2 Predator/prey ratios

From a study of 14 community foodwebs, Cohen [276] found a remarkable constancy in the ratio of the number of kinds of prey to the number of kinds of predators of approximately 3:4. This regularity has been further substantiated by Briand and Cohen [324] analysing 62 similar webs. The number of prey was roughly proportional to the number of predators, with a slope less than 1, in both fluctuating and constant environments. The use of 'kinds' of organisms rather than the actual species usually produces a biased result [274], but although this prey/predator ratio is probably too low, its constancy is interesting. Evans and Murdoch [277] found a fairly constant ratio of 1.5—2.2 between the number of species of herbivorous insect larvae and the number of entomophagous ones in grasslands during the growing season, despite a constant turnover of species. It was suggested that this reflects an underlying trophic community pattern. Cole [278] rejected this conclusion, and suggested that this ratio is maintained at a constant level merely through the statistical forces due to the random drawing of species from the total pool available. His analysis assumes that all the species listed by Evans and Murdoch are present throughout the season, which they clearly are not, because there is a species turnover. The agreement between the expected and observed numbers of herbivores found by Cole could equally arise if the insect assemblage does possess a stable trophic structure [279]. A recent study has also found that the ratio of prey species to predators is greater than 1 in arthropod assemblages from trees. Significant correlations are also apparent between the log-number of predators, parasitoids and both predators and parasitoids and in each case the log number of prey species in these assemblages (Fig. 10.5).

Another interesting finding from some of these studies is

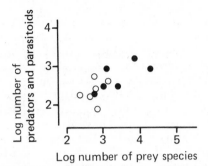

Fig. 10.5 The relationship between log abundance of predators and parasitoids and log abundance of potential prey on trees in South Africa (○) and Britain (●). (After Southwood *et al.* [250].)

that the overlap between predator species in prey use is almost always unidimensional [273, 274, 278]. This indicates that the predator trophic niche is probably one dimensional.

10.4.3 Trophic structure
The idea of stability in the whole complex trophic structure of the community was first suggested by Heatwole and Levins [280]. They subjected the data from the mangrove island recolonization study of Simberloff and Wilson [281] to closer analysis, by assigning most of the arboreal arthropods to general trophic categories or guilds. These were herbivores, scavengers, detritus feeders, woodborers, ants, predators and parasites. The number of species on six islands before and after defaunation were then tabulated guild by guild. From a total of 231 species, individual islands had a subset of between 20 and 30, and the proportion of species belonging to the guilds on each island was quickly restored although the species were different. These results suggest a remarkably stable trophic structure determined by species interaction. Simberloff [282] found no differences between these results and a corresponding pattern generated under a null hypothesis based on species drawn purely at random from the total pool. He therefore concluded that the patterns were simply chance assortments of species. However, like Cole's analysis described earlier, the null hypothesis is faulty, as it assumes that all species are available for colonization at the same time. This is not, in fact, the case, and the original data demonstrate a sequential recolonization of islands by successive trophic levels, herbivores first and carnivores later [283].

Some recent data provide further support for the existence of some form of trophic structure. Seven guilds can be recognized amongst arthropod assemblages in trees, namely phytophagous species (chewers and sap suckers), epiphyte feeders, scavengers, predators, parasitoids, ants and tourists. Comparing common, broad-leaved, tree species in Britain and South Africa, Moran and Southwood [241] have found a uniformity in the proportion of species in those guilds most closely associated with the trees, namely the phytophagous, epiphyte feeding and predatory guilds (Table 10.1). Given the wide range of total species on different trees, this proportional uniformity is impressive. Chance assortment of species might be possible, but such data indicate the existence of organized and regulated interacting assemblages of animals. Some negative interactions between the sap suckers and chewers leads to proportional uniformity in species numbers of the phytophagous guild as a whole, and this provides a hint of the sort of mechanism operating in the field [241]. Finally, in the 62 food webs studied by Briand and Cohen [324] , they subdivided species or taxa into top species (predators with no predators), intermediate species (both predators and prey) and basal species (prey with no prey). On average, the proportion of each type remained independent of the total number of species, indicating further evidence of trophic structure within communities.

10.5   Stability/diversity relationships

Traditional ecological wisdom suggests that complexity (more species and/or more interactions) implies stability (lower levels of population fluctuation, persistence or ability to recover from perturbation). However, empirical evidence is equivocal [6, 7, 105]. If complexity does confer stability on an ecosystem, one would expect populations in the tropics to be more stable than in temperate or polar regions, but there are no apparent consistent tropical—temperate differences. Studies on insect populations, for example, have shown the same average annual variability in the two zones [284]. There are also examples of simple natural systems that are stable, and of complex ones that are not. Recent studies on some freshwater ecosystems have shown that the apparently stable, more complex environments actually have a lower resilience to pertubations than less stable, simpler ones [285].

Mathematical models of ecosystems also contradict the

Table 10.1 The percentage of Arthropod species in three of the guilds recorded from broadleaved trees in South Africa (SA) and Britain (B) (adapted from Moran and Southwood [241])

Guild	Betula (Birch)		Buddleia		Erythrina	Quercus (Oak)		Robinia (Pseudoacacia)		Salix cinerea (Willow)
	SA	B	SA	B	SA	SA	B	SA	B	B
Phytophagous	21.2	26.4	20.5	26.4	18.7	24.8	22.6	24.8	23.9	26.7
Epiphyte fauna	6.1	3.9	3.6	3.4	2.3	6.7	3.2	6.7	5.5	2.5
Predators	16.5	21.1	20.1	24.2	21.7	18.8	20.9	20.0	15.6	21.7
Total number of species of all guilds	212	337	249	178	300	149		105	180	322

traditional complexity/stability notion (e.g. [286]). Randomly assembled model food webs involve three parameters, namely the number of species (S); the average connectance (C) of the web (the number of food links as a fraction of the total number possible); and the average magnitude of the interaction between linked species (i). Too great a value of C or i, tends to lead to instability in these model systems, and the larger the value of S, the more pronounced is the effect. Relatively stable and predictable environments may permit such fragile complexity, whereas relatively unstable, unpredictable environments require the dynamically robust, relatively simple community [273]. However, real communities are not randomly assembled. Through evolution, communities should evolve to be as rich and as complex as is compatible with the persistence of most populations, and to optimize use of available resources. What special features of real ecosystems can help to reconcile the observed species richness and apparent complexity of communities with the dynamical stability necessary to maintain them?

Mathematical models have identified one possible mechanism [286, 287]. Within randomly assembled foodwebs, species which interact with many others $(C$ is large), should do so weakly $(i$ is small). This is similar to the predictions of the Niche Overlap hypothesis, and suggests that diffuse competition becomes increasingly important as species diversity increases. In addition, a complex model ecosystem is robust to all types of disturbance if (a) it is a collection of self-regulating subsystems, and (b) the interactions between subsystems are weaker than the interactions within the subsystems. In other words, stability might increase with species diversity if i and/or C decrease at the same time. Do real ecosystems fit into this proposed pattern? Most insect herbivores are monophagous or oliogophagous [184], so that even in species-rich plant assemblages, relatively discrete food chains are likely. MacNaughton [288] found that both i and C decrease as the species richness (S) of vascular plants increases in 17 grassland stands (Fig. 10.6). There is a relatively constant value of SC of 4.7, despite S varying between 4 and 19. A study on 31 foodwebs involving plant—aphid—parasitoid assemblages in forest canopies also showed that C decreases as S increased. The value of SC remained around 3, whilst S varied from 3 to 60 [289]. Finally Yodzis [290] found a stable value of SC of 4 in 24 different food webs where S ranged from 8 to 64. The under-

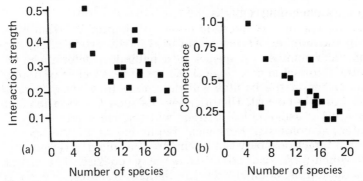

Fig. 10.6 The relationship between the average interaction strength (a) and connectance (b), and the number of species in African grassland samples. (After MacNaughton [288].)

lying mechanism keeping SC roughly constant is the tendency for these assemblages to be organized into relatively small subunits or guilds of species, with most interactions taking place within the guilds. The product SC may give some estimate of guild size [288] and as species richness increases the number of guilds may also increase (e.g. [290]). The idea of strongly interacting subunits has also been discussed with reference to intertidal communities that are based on strong trophic level interactions [178]. The occurrence of species in guilds which interact weakly with other guilds may be an important component of ecosystem organization, and it may help to reconcile the relationship between observed stability and diversity. More complete reviews of the complexity/ stability relationship at both population and community levels can be found elsewhere [325, 326].

Two other factors can reverse the general mathematical trend for complexity to decrease stability. Omnivory tends to lead to a decreased stability of model ecosystems. In nature, omnivory is not extensive and omnivores do not feed on species that are not on adjacent trophic levels [274]. Stability of communities thus seems to restrict the design of food webs. Environmental heterogeneity produces uneven population distributions, which in turn have marked effects on stability within population models. The heterogeneity of the environment is therefore a possible explanation for much of the stability of populations in the real world [238], as well as being an important factor in community structure.

10.6   Concluding remarks

Can we answer the question posed in Chapter 1: 'What controls the number of species in an area?'. To a certain extent, we
can. The number of species in a habitat is determined by
both historical and ecological factors. The number of potential niches in the habitat will determine how many species
can possibly coexist. Historical events such as speciation and
crossing of geographical barriers will determine the supply of
potential colonists. Ecological requirements of the species
and interactions between them are the main determinants
of which potential colonists actually coexist. The major
interaction is competition for limited resources, which is
the ultimate determinant of species packing and hence
species diversity. Interspecific competition promotes resource
partitioning and specialization. Coexistence is them maintained up to some limit, as a function of either the number of
discrete resources present or of the maximal tolerable niche
overlap, or both. Predation and disturbance are only proximate
controlling factors of species richness in certain situations. A
complete explanation of species diversity would need to link
biotic processes and environmental variables to explain how
patterns of immigration, speciation, competitive exclusion
and extinctions produce the observed patterns. We are clearly
not there yet.

Community attributes, such as species richness and relative
abundances are superficial indicators of community structure,
that reflect characteristics of the habitat and interactions
among the species that live there. Community structure
reflects the adaptations of each species in the community,
adaptations that are selected for in part by the activities of
coexisting species. In other words, the community is more
than simply the combination of its constituent populations.

References

[1] Menge, B. A. and Sutherland, J. P. (1976), Species diversity gradients: synthesis of the roles of predation, competition and temporal heterogeneity. *Am. Nat.*, **110**, 351–69.
Interesting and useful reading.

[2] Harper, J. L. (1980), Plant demography and ecological theory. *Oikos*, **35**, 244–53.

[3] Whittaker, R. H. (1975), *Communities and Ecosystems*, 2nd edn, Macmillan, New York.
A useful text, especially at the community level.

[4] Miles, J. (1979), *Vegetation dynamics*, (Outline Studies in Ecology), Chapman and Hall, London.
An authoritative work with an extensive bibliography.

[5] Vandermeer, J. H. (1972), Niche theory. *Ann. Rev. Ecol. Syst.*, **3**, 107–32.
A useful historical perspective and an attempt to 'standerdize' niche theory. Worth reading.

[6] McNaughton, S. J. and Wolf, L. L. (1973), *General Ecology*, Holt Reinhart and Winston, Inc., New York.

[7] May, R. M. (1981), in *Theoretical Ecology. Principles and Applications* (R. M. May ed.) 2nd edn, Blackwell Scientific Publications, Oxford.
This book introduces the reader to most of the current theoretical approaches to population and community ecology.

[8] Miller, J. C. (1980), Niche relationships among parasitic insects occurring in a temporary habitat. *Ecology*, **61**, 270–75.

[9] Pianka, E. R. (1978), *Evolutionary Ecology* 2nd edn, Harper and Row, New York.
Highly recommended text.

[10] Ricklefs, R. E. (1980), *Ecology*, 2nd edn, Nelson, Walton on Thames.
The best general ecology text currently available.

[11] Whittaker, R. H. (1972), Evolution and measurement of species diversity. *Taxon*, 21, 213—51.
Heavy but essential reading.

[12] Peet, R. K. (1974), The measurement of species diversity. *Ann. Rev. Ecol. Syst.*, 5, 285—307.

[13] Pianka, E. R. (1966), Latitudinal gradients in species diversity: A review of concepts. *Am. Nat.*, 100, 33—34.

[14] Schall, J. J. and Pianka, E. R. (1978), Geographical trends in numbers of species. *Science*, 201, 679—86.

[15] Fischer, A. G. (1960), Latitudinal variations in organic diversity. *Evolution*, 14, 64—81.
A classic contribution with many examples.

[16] Rabenold, K. N. (1979), A reversed latitudinal diversity gradient in avian communities of Eastern deciduous forests. *Am. Nat.*, 114, 275—86.

[17] Parsons, P. A. and Bock, I. R. (1979), Latitudinal species diversities in Australian endemic *Drosophila*. *Am. Nat.*, 114, 213—20.

[18] Richards, P. W. (1979), *Speciation in Tropical Environments*, Academic Press, New York.

[19] Patten, B. C. and Auble, G. T. (1981), System theory of the ecological niche. *Am. Nat.*, 117, 893—922.
Pushes the use of mathematics in ecology too far.

[20] Hutchinson, G. E. (1975), A theme by Robert MacArthur, in *Ecology and Evolution of Communities* (M. L. Cody and J. M. Diamond eds), Harvard University Press, Cambridge, pp. 492—521.
This book presents a unique collection of papers presented at a symposium in memory of Robert MacArthur. Highly recommended.

[21] Grinnell, J. (1917), Field tests of theories concerning distribution control. *Am. Nat.*, 51, 115—28.

[22] Elton, C. (1927), *Animal Ecology*. Sidgewick and Jackson, London.

[23] Hardin, G. (1960), The competitive exclusion principle. *Science*, 131, 1292—297.

[24] Odum, E. P. (1971), *Fundamentals of Ecology*, 2nd edn, W. B. Saunders, Philadelphia.

[25] Hutchinson, G. E. (1958), Concluding remarks. *Cold Spring Harbor Symp. Quant. Biol.*, 22, 415—27.

[26] Roughgarden, J. (1974), Niche width: biogeographic patterns among *Anolis* Lizard populations. *Am. Nat.*, 108, 429—42.

[27] Pianka, E. R. (1974), Niche overlap and diffuse competition. *Proc. Nat. Acad. Sci.*, 71, 2141—145.

[28] MacArthur, R. H. (1972), *Geographical Ecology. Patterns in the Distribution of Species.* Harper and Row, New York.
One of the milestones in the ecological literature.

[29] Southwood, T. R. E. (1978), *Ecological Methods*, 2nd edn,

Chapman and Hall, London.
An outstanding review of methods covering all branches of animal ecology.

[30] Van Valen, L. (1968), Morphological variation and width of ecological niche. *Am. Nat.*, **99**, 377–90.

[31] Levins, R. (1968), *Evolution in Changing Environments*, Princetown University Press, New Jersey.

[32] Hurlbert, S. H. (1978), The measurement of niche overlap and some relatives. *Ecology*, **59**, 67–77.

[33] Feinsinger, P., Spears, E. E. and Poole, R. W. (1981), A simple measure of niche breadth. *Ecology*, **62**, 27–32.

[34] Smith, E. P. and Zaret, T. M. (1982), Bias in estimating niche overlap. *Ecology*, **63**, 1248–253.
A useful, critical review of niche overlap measures.

[35] Slobodchikoff, C. N. and Schulz, W. C. (1980), Measures of niche overlap. *Ecology*, **61**, 1051–55.

[36] Abrams, P. (1980), Some comments on measuring niche overlap. *Ecology*, **61**, 44–9.
A critical examination of the relationship between niche overlap and competition.

[37] Davies, R. W., Wrona, F. J. and Linton, L. (1979). A seriological study of prey selection by *Helobdella stagnalis* (Hirudinoidea). *J. Anim. Ecol.*, **48**, 181–94.

[38] Rusterholz, K. A. (1981), Competition and the Structure of an avian foraging guild. *Am. Nat.*, **118**, 173–90.

[39] Clapham, W. B. (1973), *Natural Ecosystems*, MacMillan, New York.

[40] Pyke, G. H., Pulliam, H. R. and Charnov, E. L. (1977), Optimal Foraging: a selective review of theory and tests. *Quart. Rev. Biol.*, **52**, 138–55.
A concise review of optimal foraging theory.

[41] Painka, E. R. (1975), Niche relations of desert lizards, in *Ecology and Evolution of Communities* (M. L. Cody, and J. M. Diamond, eds), Harvard University Press, Cambridge, pp. 292–314.

[42] Cody, M. L. (1975), Towards a theory of continental species diversities, in *Ecology and Evolution of Communities* (M. L. Cody, and J. M. Diamond, eds), Harvard University Press. Cambridge. pp. 214–57.

[43] Karr, J. R. and James, F. C. (1975), Ecomorphological configurations and convergent evolution, in *Ecology and Evolution of Communities* (M. L. Cody and J. M. Diamond, eds) Harvard University Press, Cambridge, pp. 258–91.

[44] Abramsky, Z. and Sellah, C. (1982), Competition and the role of habitat selection in *Gerbillus allenbyi* and *Meriones tristrami*: A removal experiment. *Ecology*, **63**, 1242–247.

[45] Diamond, J. M. (1978), Niche shifts and the rediscovery of interspecific competition. *Am. Scientist*, **66**, 322–31.

[46] O'Connor, R. J., Boaden, P. J. S. and Seed, R. (1975), Niche

breadth in Bryozoa and a test of competition theory. *Nature*, **256**, 307—09.
A natural example of the effects of intra- and interspecific competition on niche width.

[47] Charnov, E. L. (1976). Optimal foraging, the Marginal Valve theorm. *Theor. Popul. Biol.*, **9**, 129—36.

[48] Pimm, S. and Pimm, J. W. (1982), Resource use, competition, and resource availability in Hawaiian honeycreepers. *Ecology*, **63**, 1468—480.

[49] Giller, P. S. (1980), The control of handling time and its effect on the foraging strategy of a heteropteran predator, *Notonecta*. *J. Anim. Ecol.*, **49**, 699—712.

[50] Nilsson, N. (1965), Food segregation between salmonid species in North Sweden. *Rep. inst. Freshw. Res. Drottingholm*, **46**, 58—78.

[51] Pontin, A. J. (1982), *Competition and Coexistence of Species.* Pitman, London.
An interesting and readable account but lacks many of the more recent findings.

[52] Hassell, M. P. (1976), *The Dynamics of Competition and Predation.* Edward Arnold, Studies in Biology, No. 72.

[53] Lock, M. A. and Reynoldson, T. B. (1976), The role of interspecific competition in the distribution of two stream dwelling triclads, *Crenobia alpina* (Dana) and *Polycelis felina* (Dalyell) in North Wales *J. Anim. Ecol.*, **45**, 581—92.

[54] Philips, D. L. and MacMahon, J. A. (1981), Competition and spacing patterns in desert shrubs. *J. Ecol.*, **69**, 97—117.

[55] Black, R. (1979), Competition between intertidal limpets: an intrusive niche on a steep resource gradient. *J. Anim. Ecol.*, **48**, 401—11.

[56] Selander, R. K. (1966), Sexual dimorphism and differential niche utilisation in birds. *The Condor*, **68**, 113—51.

[57] Diamond, J. M. (1973), Distributional ecology of New Guinea birds. *Science*, **179**, 759—69.
An important contribution.

[58] Brown, J. H. (1975), Geographical ecology of desert rodents. in *Ecology and Evolution of Communities*, (Cody, M. L. and Diamond, J. M., eds), Harvard University Press, Cambridge, pp. 315—41.

[59] Glass, G. E. and Slade, N. A. (1980), The effect of *Sigmondon hispidus* on spatial and temporal activity of *Microtus ochrogastor:* evidence for competition. *Ecology*, **61**, 358—70.

[60] Grant, J. (1981), Dynamics of competition among estuarine sand-burrowing amphipods. *J. Exp. Mar. Biol. Ecol.*, **49**, 255—67.

[61] Gorman, M. (1979), *Island Ecology*. Chapman and Hall (Outline Studies in Ecology), London.

[62] Nilsson, S. G. (1977), Density compensation among birds

breeding on small islands in a south Swedish lake. *Oikos*, **28**, 170—76.

[63] Feinsinger, P. and Swarm, L. (1982), Ecological release, seasonal variation in food supply and the hummingbird *Amazila tobaci* on Trinidad and Tobago. *Ecology*, **63**, 1574—587.

[64] Davidson, D. W., Brown, J. H. and Inouye, R. S. (1980), Competition and the structure of granivore communities. *Bioscience*, **30**, 233—38.
An interesting example of resource partitioning and competition across taxonomic boundaries.

[65] Fowler, N. (1982), Competition and coexistence .in a North Carolina grassland, III. Mixtures of component species. *J. Ecol.* **70**, 77—92.
A good example of experimental methods designed to investigate possible competitive interactions.

[66] Findley, J. S. (1976), The structure of bat communities. *Am. Nat.*, **110**, 129—39.

[67] Steiner, W. W. M. (1980), On niche width and genetic variation, parametrical problems, and gene pool differentiation in *Drosophila*. *Am. Nat.*, **115**, 596—99.

[68] Rusterholz, K. A. (1981), Niche overlap among foliage-gleaning birds: support for Pianka's Niche Overlap hypothesis. *Am. Nat.*, **117**, 395—99.

[69] Shugart, H. H. and Blaylock, B. G. (1973), The Niche Variation hypothesis: an experimental study with *Drosophila* populations. *Am. Nat.*, **107**, 575—79.

[70] Soule, M. and Stewart, B. R. (1970), The Niche Variation hypothesis: a test and alternatives. *Am. Nat.*, **104**, 85—97.

[71] Van Valen, L. and Grant, P. R. (1970), Variation and niche width re-examined. *Am. Nat.*, **104**, 589—90.

[72] Jackson, J. B. C. (1979), Overgrowth competition between encrusting cheilostome ectoprocts in a Jamaican cryptic reef environment. *J. Anim. Ecol.*, **48**, 805—24.
A demonstration of competition in the field.

[73] Burrough, R. J., Bregazzi, P. R. and Kennedy, C. R. (1979), Interspecific dominance amongst three species of coarse fish in Slapton Ley, Devon. *J. Fish Biol*, **15**, 535—44.

[74] Miracle, M. R. (1974), Niche structure in freshwater zooplankton: A principal components approach. *Ecology*, **55**, 1306—316.

[75] Harper, J. L. (1977), *Population Biology of Plants*, McGraw-Hill, London.
An authoritative text including sections on community structure and diversity, competition and the role of predation on vegetation.

[76] Fonteyn, P. J. and Mahall, B. E. (1981), An experimental analysis of structure in a desert plant community. *J. Ecol.*, **69**, 883—96.

[77] Fowler, N. (1981), Competition and coexistence in a North Carolina grassland, II. The effects of the experimental removal of species. *J. Ecol.*, **69**, 843—54.

[78] Putwain, P. D. and Harper, J. L. (1970), Studies in the dynamics of plant populations, III. The influence of associated species on populations of *Rumex acetosa* L. and *R. acetosella* L. in grassland. *J. Ecol.*, **58**, 251—64.

[79] Cideciyan, M. A. and Malloch, A. J. C. (1982), Effects of seed size on the germination, growth and competitive ability of *Rumex crispus* and *Rumex obtusifolius*. *J. Ecol.*, **70**, 227—32.

[80] Grant, P. R. (1972), Interspecific competition among rodents. *Ann. Rev. Ecol. Syst.*, **3**, 79—106.
Describes many experimental studies of competitive interaction in a two-species system.

[81] Hairston, N. G. (1981), An experimental test of a guild: Salamander competition. *Ecology*, **62**, 65—72.

[82] Shorten, M. (1954), *Squirrels*. Collins New Naturalist series, London.

[83] Connell, J. H. (1975), Some mechanisms producing structure in natural communities, in *Ecology and Evolution of Communities* (Cody, M. L. and Diamond, J. M., eds), Harvard University Press, Cambridge, pp. 460—91.

[84] Joseph Wright, S. (1981), Extinction-mediated competition: The *Anolis* lizards and insectivorous birds of the West Indies. *Am. Nat.*, **117**, 181—92.

[85] Diamond, J. M. (1979), Community structure: is it random, or is it shaped by species differences and competition?, In *Population Dynamics* (Anderson, R. M., Turner, B. D. and Taylor, L. R. eds), pp. 165—82, Blackwell Scientific Publications, Oxford.
Uses examples from islands to discuss the nature of community organization. The book as a whole presents a series of informative papers on population and community ecology written by some of the more eminent workers in the field.

[86] Porter, J. H. and Dueser, R. D. (1982), Niche overlap and competition in an insular small mammal fauna: a test of the niche overlap hypothesis. *Oikos*, **39**, 228—36.

[87] Morse, D. H. (1977), Resource partitioning in Bumblebees: the role of behavioural factors. *Science*, **197**, 678—80.

[88] Hairston, N. (1951), Interspecies competition and its probable influence upon vertical distribution of Appalachian Salamanders of the genus *Plethodon*, *Ecology*, **32**, 266—74.

[89] Krzysik, A. J. (1979), Resource allocation, coexistence, and the niche structure of a streambank salamander community. *Ecol. Monogr.*, **49**, 173—94.

[90] Schoener, T. W. (1975), Presence and absence of habitat shift in some widespread lizard species. *Ecol. Mongr.*, **45**, 232—58.

[91] Carpenter, F. L. (1979), Competition between hummingbirds and insects for nectar. *Amer. Zool.*, **19**, 1105—114.

[92] Grant, P. R. (1972), Convergent and divergent character displacement. *Biol. J. Linn. Soc.*, 4, 39—68.
Critical review of evidence for morphological character displacement, but now out of date.

[93] Huey, R. B. and Pianka, E. R. (1974), Ecological character displacement in a lizard. *Am. Zool.*, 14, 1127—136.

[94] Fenchel, T. (1975), Character displacement and coexistence in mud snails (Hydrobiidae). *Oecologia*, 20, 19—32.

[95] Levinton, J. S. (1982), The body size—prey size hypothesis: the adequacy of body size as a vehicle for character displacement. *Ecology*, 63, 869—72.

[96] Van Zant, T., Poulson, T. L. and Kane, T. C. (1978), Body-size differences in Carabid cave beetles. *Am. Nat.*, 112, 229—34.

[97] Barr, T. C. and Crowley, P. H. (1981), Do cave carabid beetles really show character displacement in body size? *Am. Nat.*, 117, 363—71.

[98] Wilson, D. (1975), The adequacy of body size as a niche difference. *Am. Nat.*, 109, 769—84.

[99] Slatkin, M. (1980), Ecological character displacement. *Ecology*, 61, 163—77.

[100] Smith, J. N. M., Grant, P. R., Grant, B. R., Abbott, I. J. and Abbott, L. K. (1978), Seasonal variation in feeding habitats of Darwin's ground finches. *Ecology*, 59, 1137—150.
Examines the effects of variation in food abundance on food partitioning. A superb study in the best traditions of the field ecologist.

[101] Zaret, T. M. and Rand, A. S. (1971), Competition in tropical stream fishes: support for the competitive exclusion principle. *Ecology*, 52, 336—42.
One of the best studies to date.

[102] Brown, J. H. and Lieberman, G. A. (1973), Resource utilisation and coexistence of seed-eating desert rodents in sand dune habitats. *Ecology*, 54, 788—97.

[103] Cody, M. L. (1973), Coexistence, coevolution and convergent evolution in seabird communities. *Ecology*, 54, 31—43. *A classic example of resource partitioning.*

[104] Emmons, L. H. (1980), Ecology and resource partitioning among nine species of African rain forest squirrels. *Ecol. Monog.*, 50, 31—54.

[105] Davies, R. W., Wrona, F. J., Linton, L. and Wilkialis, J. (1981), Inter- and intra-specific analyses of the food niche of two sympatric species of Erpobdellidae (Hirudinoidea) in Alberta, Canada. *Oikos*, 37, 105—11.

[106] Fox, B. J. (1981), Niche parameters and species richness. *Ecology*, 62, 1415—425.

[107] Ricklefs, R. E., Cochran, O. and Pianka, E. R. (1981), A morphological analysis of the structure of communities of lizards

in desert habitats. *Ecology*, **62**, 1474—483.

[108] M'Closkey, R. T. (1978), Niche separation and assembly in four species of Sonoran desert rodents. *Am. Nat.*, **112**, 683—94.

[109] Bernstein, R. A. and Gobbel, M. (1979), Partitioning of space in communities of ants. *J. Anim. Ecol.*, **48**, 931—42.

[110] Abrams, P. A. (1977), Density independent mortality and interspecific competition: A test of Pianka's niche overlap hypothesis. *Am. Nat.*, **111**, 539—52.

[111] MacArthur, R. H. (1965), Patterns of species diversity. *Biol. Rev.*, **40**, 510—33.

[112] Abrams, P. (1975), Limiting similarity and the form of the competition coefficient. *Theor. Pop. Biol.*, **8**, 356—75.

[113] Roughgarden, J. (1976), Resource partitioning among competing species: a co-evolutionary approach. *Theor. Pop. Biol.*, **9**, 388—424.

[114] MacArthur, R. H. and Levins, R. (1967), The limiting similarity, convergence and divergence of coexisting species. *Am. Nat.*, **101**, 377—85.

[115] Pianka, E. R. (1969), Sympatry of desert lizards (*Ctenotus*) in Western Australia. *Ecology*, **50**, 1012—30.

[116] Schoener, T. W. (1968), The *Anolis* lizards of Bimini: resource partitioning in a complex fauna. *Ecology*, **49**, 704—26.

[117] Hespenheide, H. A. (1971), Food preference and the extent of overlap in some insectivorous birds, with special reference to the Tyrannidae. *Ibis*, **113**, 59—72.

[118] McNab, B. K. (1971), The structure of tropical bat faunas. *Ecology*, **52**, 352—56.

[119] Pearson, D. L. and Mury, E. J. (1979), Character divergence, and convergence among tiger beetles (Coleoptera; Cicindelidae). *Ecology*, **60**, 557—66.

[120] Uetz, G. W. (1977), Coexistence in a guild of wandering spiders. *J. Anim. Ecol.*, **46**, 531—42.

[121] Hutchinson, G. E. (1959), Homage to Santa Rosalia or Why are there so many kinds of animals? *Am. Nat.*, **93**, 145—59.

[122] Horn, H. S. and May, R. M. (1977), Limits to similarity among coexisting competitors, *Nature*, **270**, 660—61.

[123] Schoener, T. W. (1965), The evolution of bill size differences among sympatric congeneric species of birds. *Evolution*, **19**, 189—213.

[124] Lawton, J. H. and Strong, D. R. (1981), Community patterns and competition in folivorous insects. *Am. Nat.*, **118**, 317—38. *Well-constructed arguments to counter the belief that general community patterns found in these insect assemblages are attributable to interspecies competition.*

[125] Maiorana, V. C. (1978), An explanation of ecological and developmental constants. *Nature*, **273**, 375—77.

[126] Roth, V. L. (1981), Constancy in the size ratios of sympatric species. *Am. Nat.*, **118**, 394—404.

[127] Turner, M. and Polis, G. (1979), Patterns of co-existence in a guild of raptorial spiders. *J. Anim. Ecol.*, **48**, 509—20.

[128] Davies, N. B. (1977), Prey selection and the search strategy of the spotted flycatcher (*Muscicapa striata*): A field study on optimal foraging. *Anim. Behav.*, **25**, 1016—33.

[129] Hespenheide, H. (1973), Ecological inferences from morphological data. *Ann. Rev. Ecol. Syst.*, **4**, 213—29.

[130] Enders, F. (1974), Vertical stratification in orb-web spiders (Araneidae, Araneae) and a consideration of other methods of coexistence. *Ecology*, **55**, 317—28.

[131] Edington, J. M. and Edington, M. A. (1972), Spatial patterns and habitat partition in the breeding birds of an upland wood. *J. Anim. Ecol.*, **41**, 331—57.
An excellent field study to investigate mechanisms of species coexistence.

[132] Giller, P. S. and McNeill, S. (1981), Predation strategies, resource partitioning and habitat selection in *Notonecta* (Hemiptera/ Heteroptera). *J. Anim. Ecol.*, **50**, 789—808.

[133] Alatalo, R. V. and Alatalo, R. H. (1979), Resource partitioning among a flycatcher guild in Finland. *Oikos*, **33**, 46—54.

[134] Anderson, G. R. V., Ehrlich, A. H., Ehrlich, P. R., Roughgarden, J. D., Russell, B. C. and Talbot, F. H. (1981), The community structure of coral fishes. *Am. Nat.*, **117**, 476—95.

[135] Cody, M. L. (1968), On the methods of resource division in grassland bird communities. *Am. Nat.*, **102**, 107—47.

[136] Schoener, T. W. (1974), Resource partitioning in ecological communities. *Science*, **185**, 27—39.
A classic paper. Essential reading.

[137] Connell, J. H. (1978), Diversity in tropical rain forests and coral reefs. *Science*, **199**, 1302—310.
Attacks the long-held view that the ecological diversity in the tropics is the result of stable conditions.

[138] Harper, J. L. (1969), The role of predation in vegtational diversity. *Brookhaven Symp. Biol.*, **22**, 48—62.

[139] Carter, R. and Prince, S. (1981), Epidemic models used to explain biogeographical distribution limits. *Nature*, **293**, 644—45.

[140] Whittaker, R. (1965), Dominance and diversity in land plant communities. *Science*, **147**, 250—60.
A useful review of species abundance distributions in plant assemblages.

[141] Grime, J. P. (1979), Competition and the struggle for existence, in *Population Dynamics* (Anderson, R. M., Turner, B. D. and Taylor, L. R. eds), Blackwell Scientific Publications, Oxford, pp. 123—40.

[142] McCown, R. L. and Williams, W. A. (1968), Competition for nutrients and light between the annual grassland species, *Bromus mollis* and *Erodium botrys*. *Ecology*, **49**, 981—90.

[143] Harper, J. L. and Bell, A. D. (1979), The population dynamics of growth form in organisms with modular construction, in *Population Dynamics* (Anderson, R. M., Turner, B. D. and Taylor, L. R. eds), Blackwell Scientific Publications, Oxford, pp. 29—52.

[144] Grime, J. P. (1977), Evidence for the existence of three primary strategies in plants and its relevance to ecological and evolutionary theory. *Am. Nat.*, 111, 1169—194.

[145] Hespenheide, H. A. (1975), Prey characteristics and predator niche width, in *Ecology and Evolution of Communities* (Cody, M. L. and Diamond, J. M. eds), Harvard University Press, Cambridge, pp. 158—80.

[146] Williamson, P. (1971), Feeding ecology of the red-eyed vireo (*Vireo olivaceous*) and associated foliage gleaning birds. *Ecol. Monogr.*, 41, 129—52.

[147] Morse, D. H. (1968), A quantitative study of foraging of male and female sprucewoods warblers. *Ecology*, 49, 779—84.

[148] Christian, D. P. (1980), Vegatative cover, water resources and microdistributional patterns in a desert rodent community. *J. Anim. Ecol.*, 49, 807—16.

[149] Onyekwelu, S. S. and Harper, J. L. (1979), Sex ratio and niche differentiation in spinach (*Spinacia oleracea* L.). *Nature*, 282, 609—11.

[150] Cox, P. A. (1981), Niche partitioning between sexes of dioecious plants. *Am. Nat.*, 117, 295—307.

[151] Smigel, B. W. and Rosenzweig, M. L. (1974), Seed selection in *Dipodomys merriami* and *Perognathus penicillatus*. *Ecology*, 55, 329—39.

[152] McClure, M. S. and Price, P. W. (1976), Ecotype characteristics of coexisting *Erythroneura* leafhoppers (Homoptera cicadellidae) on American Sycamore. *Ecology*, 57, 929—40.

[153] Pianka, E. R. (1973), The structure of lizard communities. *Ann. Rev. Ecol. Syst.*, 4, 53—74.

[154] Usher, M. B., Davis P. R., Harris, J. R. W. and Longstaff, B. C. (1979), A profusion of species? Approaches towards understanding the dynamics of the populations of the microarthropods in decomposer communities, in *Population Dynamics*, (Anderson, R. M., Turner, B. D. and Taylor, L. R. eds) Blackwell Scientific Publications, Oxford, pp. 359—384.

[155] White, J. (1980), Resource partitioning by ovipositing cicadas. *Am. Nat.*, 115, 1—28.

[156] Gillis, J. E. and Possai, K. W. (1983), Thermal niche partitioning in the grasshoppers *Arphia conspersa* and *Trimerotropis suffusa* from a montane habitat in central Colorado. *Ecol. Ent.*, 8, 155—61.

[157] Stiling, P. (1980), Competition and coexistence among *Eupteryx* leafhoppers (Hemiptera: Cicadellidae) occurring on stinging nettles (*Urtica dioica*). *J. Anim Ecol.*, 49, 793—805.

[158] Wolda, H. and Fisk, F. (1981), Seasonality of tropical insects. II. *Blattaria* in Panama. *J. Anim. Ecol.*, 50, 827—38.

[159] Plowman, K. P. (1981), Resource partitioning by two New Guinea rainforest ants. *J. Anim. Ecol.*, **50**, 903—17.

[160] Hildrew, A. and Edington, J. (1979), Factors facilitating co-existence of hydropsychid caddis larvae (Trichoptera) in the same river system. *J. Anim. Ecol.*, **48**, 557—76.

[161] Giller, P. S. (1982), Locomotory efficiency in the predation strategies of the British *Notonecta* (Hemiptera, Heteroptera). *Oecologia*, **52**, 273—77.

[162] Sandercock, G. A. (1967), A study of selected mechanisms for the coexistence of *Diaptomus* spp. in Clarke Lake, Ontario. *Limnol. Oceanogr.*, **12**, 97—112.

[163] Smith, C. C. and Balda, R. P. (1979), Competition among insects, birds and mammals for confier seeds. *Amer. Zool.*, **19**, 1065—83.
Discusses the coexistence of many distantly related taxa exploiting an easily delimited resource.

[164] Townsend, C. R. and Hildrew, A. G. (1979), Resource partitioning by two freshwater invertebrate predators with contrasting foraging strategies. *J. Anim. Ecol.*, **48**, 909—20.

[165] Roughgarden, J. and Feldman, M. (1975). Species packing and predation pressure. *Ecology*, **56**, 489—92.

[166] Hassell, M. P. (1979), The dynamics of predator—prey interactions: polyphagous predators, competing predators and hyperparasitoids, in *Population Dynamics* (Anderson, R. M., Turner, B. D. and Taylor, L. R. eds), Blackwell Scientific Publications, Oxford, pp. 282—306.

[167] Glasser, J. W. (1979), The role of predation in shaping and maintaining the structure of communities. *Am. Nat.*, **113**, 631—41.
Suggests a novel explanation for the relationship between predation and prey species diversity, but it is not entirely convincing.

[168] Ayala, F. and Cambell, C. (1974), Frequency-dependent selection. *Ann. Rev. Ecol. Syst.*, **5**, 115—38.

[169] Lee, T. D. and Bazzaz, F. A. (1980), Effects of defoliation and competition on growth and reproduction in the annual plant *Abutilon theophrasti. J. Ecol.*, **68**, 813—22.

[170] Hairston, N. G., Smith, F. E. and Slobodkin, L. B. (1960), Community structure, population control and competition. *Am. Nat.*, **94**, 421—25.

[171] Hodkinson, I. D. and Hughes, M. K. (1982), *Insect Herbivory.* Chapman and Hall, (Outline Studies in Ecology), London.

[172] Whittaker, J. B. (1979), Invertebrate grazing, competition and plant dynamics, in *Population Dynamics* (Anderson, R. M., Turner, B. D. and Taylor, L. R. eds), Blackwell Scientific Publications, Oxford, pp. 207—22.
A useful review on the effect of grazing by invertebrates on competition between their food plants.

[173] Bentley, S., Whittaker, J. B. and Malloch, A. J. C. (1980), Field experiments on the effects of grazing by a Chrysomelid beetle

(*Gastrophysa viridula*) on seed production and quality in *Rumex obtusifolius* and *Rumex crispus. J. Ecol.*, **68**, 671—74.

[174] Boorman, L. A. and Fuller, R. M. (1982), Effects of added nutrients on dune swards grazed by rabbits. *J. Ecol.*, **76**, 345—55.

[175] Inouye, R. S., Byers, G. S. and Brown, J. H. (1980), Effects of predation and competition on survivorship, fecundity, and community structure of desert annuals. *Ecology*, **61**, 1344—351.

[176] Hay, M. E. (1981), Herbivory, algal distribution and the maintenance of between-habitat diversity on a tropical fringing reef. *Am. Nat.*, **118**, 520—40.
A good demonstration of the use of exclusion and transplatation experiments.

[177] Robles, C. D. and Cubit, J. (1981), Influence of biotic factors in an upper intertidal community: Dipteran larvae grazing on algae. *Ecology*, **62**, 1536—547.

[178] Paine, R. T. (1980), The third Tansley Lecture. Food webs, interaction strength and community infrastructure. *J. Anim. Ecol.*, **49**, 667—86.

[179] Estes, J. A. and Palmisano, J. F. (1974), Sea otters: their role in structuring near shore communities. *Science*, **185**, 1058—60.

[180] Janzen, D. H. (1971), Seed predation by animals. *Ann., Rev. Ecol. Syst.*, **2**, 465—91.

[181] Watt, A. S. (1981), Further observations on the effects of excluding rabbits from grassland A in East Anglian breckland: The pattern of change and factors affecting it (1936—73). *J. Ecol.*, **69**, 509—36.

[182] Moss, R., Watson, A. and Ollason, J. (1982), *Animal Population Dynamics.* Chapman and Hall, (Outline Studies in Ecology), London.

[183] Anderson, R. M. (1979), The influence of parasitic infection on the dynamics of host population growth, in *Population Dynamics* (Anderson, R. M., Turner, B. D. and Taylor, L. R. eds), Blackwell Scientific Publications, Oxford, pp. 245—82.

[184] Lawton, J. H. and McNeill, S. (1979), Between the devil and the deep blue sea: on the problem of being a herbivore, in *Population Dynamics* (Anderson, R. M., Turner, B. D. and Taylor, L. R. eds), Blackwell Scientific Publications, Oxford, pp. 223—44.
A thought-provoking review of some of the mechanisms determining the characteristic levels of abundance of plant-feeding insects.

[185] Caughley, G. and Lawton, J. H. (1981), Plant—herbivore systems, in *Theoretical Ecology. Principles and Applications* (May, R. M. ed.), 2nd edn, Blackwell Scientific Publications, Oxford, pp. 132—66.

[186] Faeth, S. H., Mopper, S. and Simberloff, D. (1981), Abundances and diversity of leaf-mining insects on three oak host species: effects of host—plant phenology and nitrogen content of leaves. *Oikos*, **37**, 238—51.

[187] Neill, W. E. (1975), Experimental studies of microcrustacean competition, community composition and efficiency of resource utilization. *Ecology*, **56**, 809—26.
An excellent example of laboratory experiments designed to investigate competition and predation effects on species co-existence.

[188] Moss, B. (1980), *Ecology of Freshwaters*. Blackwell Scientific Publications, Oxford.

[189] Fulton, R. S. (1982), Preliminary results of an experimental study of the effects of mysid predation on estuarine zooplankton community structure. *Hydrobiolgia*, **93**, 79—84. ·

[190] Paine, R. T. (1966), Food web complexity and species diversity. *Am. Nat.*, **100**, 65—74.

[191] Porter, J. W. (1972), Predation by *Acanthaster* and its effects on coral species diversity. *Am. Nat.*, **106**, 487—92.

[192] Russ, G. R. (1980), Effects of predation by fishes, competition and structual complexity of the substratum on the establishment of a marine epifaunal community.*J. Exp. Mar. Biol. Ecol.*, **42**, 55—61.

[193] Addicott, J. F. (1974), Predation and prey community structure: an experimental study of the effect of mosquito larvae on protozoan communities in pitcher plants. *Ecology*, **55**, 475—92.

[194] Allan, J. D. (1982), The effects of reduction in trout density on the invertebrate community of a mountain stream. *Ecology*, **63**, 1444—455.

[195] Connell, J. H. (1980), Diversity and the coevolution of competitors or the ghost of competition past. *Oikos*, **35**, 131—38.

[196] Grime, J. P. (1973), Competitive exclusion in herbaceous vegetation. *Nature*, **242**, 344—47.

[197] Hassell, M. P., Lawton, J. H. and Beddington, J. R. (1977), Sigmoid functional responses by invertebrate predators and parasitoids. *J. Anim. Ecol.*, **46**, 249—62.

[198] Thompson, S. D. (1982), Structure and species composition of desert heteromyid rodent species assemblages: effects of a simple habitat manipulation. *Ecology*, **63**, 1313—321.

[199] Hildrew, A. G. and Townsend, C. R. (1977), The influence of substrate on the functional response of *Plectronemia conspersa* (Curtis) larvae (Trichoptera, Polycentropodidae). *Oecologia*, **31**, 21—6,

[200] Rosenzweig, M. L. (1975), On continental steady states of species diversity, in *Ecology and Evolution of Communities* (Cody, M. L. and Diamond, J. M. eds.), Harvard University Press, Cambridge, pp. 121—41.

[201] Simberloff, D. S. (1974), Equilibrium theory of island biogeography and ecology. *Ann. Rev. Ecol. Syst.*, **5**, 161—82.
A useful review of MacArthur and Wilson's Equilibrium Theory.

[202] Diamond, J. M. (1974), Colonisation of exploded volcanic islands by birds: the supertramp strategy. *Science*, **184**, 803—6.

[203] Diamond, J. M. and May, R. M. (1981), Island biogeography and the design of nature reserves, in *Theoretical Ecology. Principles and Applications* (May, R. M. ed.), Blackwell Scientific Publications, Oxford, pp. 228–52.
Presents a useful list of species–area exponent values (Z) for groups of plants and animals in various parts of the world.

[204] Schoener, T. W. and Schoener, A. (1983), Distribution of vertebrates on some very small islands II. Patterns in species number. *J. Anim. Ecol.*, **52**, 237–62.

[205] Krebs, C. J. (1978), *Ecology. The Experimental Analysis of Distribution and Abundance* (2nd edn), Harper and Row, New York.
An excellent textbook for both population and community ecology. Presents some useful species diversity case studies.

[206] Patrick, R. (1967), The effect of invasion rate, species pool and size of area on the structure of the diatom community. *Proc. Nat. Acad. Sci.*, **58**, 1335–342.

[207] Schoener, A. and Schoener, T. W. (1981), The dynamics of the species–area relation in marine fouling systems: 1. Biological correlates of changes in the species area slope. *Am. Nat.*, **118**, 339–60.

[208] Connor, E. F. and McCoy, E. D. (1979), The statistics and biology of the species–area relationship. *Am. Nat.*, **113**, 791–833.
An extensive review of the subject.

[209] May, R. M. (1975), Patterns of species abundance and diversity, in *Ecology and Evolution of Communities* (Cody, M. L. and Diamond, J. M. eds) Harvard University Press, Cambridge, pp. 81–120.

[210] Sugihara, G. (1980), Minimal community structure: an explanation of species abundance patterns. *Am. Nat.*, **116**, 770–87.

[211] Armesto, J. J. and Contreras, L. C. (1981), Saxicolous lichen communities: nonequilibrium systems? *Am. Nat.*, **118**, 597–604.

[212] Terborgh, J. (1973), On the notion of favourableness in plant ecology. *Am. Nat.*, **107**, 481–501.

[213] Diamond, J. M. (1975), Assembly of species communities, in *Ecology and Evolution of Communities* (Cody, M. L. and Diamond, J. M. eds), Harvard University Press, Cambridge, pp. 342–444.
Explores differences in community structure based on the hypothesis that, through diffuse competition, the component species of an island community are selected and coadjusted in their niche and abundances, so as to fit with each other and to resist invaders.

[214] Schoener, T. W. and Schoener, A. (1983), Distribution of vertebrates on some very small islands. I. Occurrence sequences of individual species. *J. Anim. Ecol.*, **52**, 209–36.

[215] Terborgh, J. W. and Faaborg, J. (1980), Saturation of bird communities in the West Indies. *Am. Nat.*, **116**, 178—95.

[216] Connor, E. F., Faeth, S. H., Simberloff, D. and Opler, P. A. (1980), Taxonomic isolation and the accumulation of herbivorous insects: a comparison of introduced and native trees. *Ecol. Ent.*, **5**, 205—11.

[217] Strong, D. R. and Levin, D. A. (1979), Species richness of plant parasites and growth form of their hosts. *Am. Nat.*, **114**, 1—22.

[218] Claridge, M. F. and Wilson, M. R. (1982), Insect herbivore guilds and species—area relationships: leafminers on British trees. *Ecol. Ent.*, **7**, 19—30.

[219] Kuris, A. M., Blaustein, A. R. and Alio, J. J. (1980), Hosts as islands. *Am. Nat.*, **116**, 570—86.

[220] Rey, J. R., McCoy, E. D. and Strong, D. R. (1981), Herbivore pests, habitat islands and the species—area relation. *Am. Nat.*, **117**, 611—22.

[221] Lawton, J. H., Cornell, H., Dritschilo, W. and Hendrix, S. D. (1981), Species as islands: comments on a paper by Kuris *et al. Am. Nat.*, **117**, 623—27.

[222] Tuomi, J., Niemela, P. and Mannila, R. (1981), Leaves as islands: interactions of *Scolioneura betuleti* (Hymenoptera) miners in birch leaves. *Oikos*, **37**, 146—52.

[223] Horn, H. (1975), Markovian processes in forest succession, in *Ecology and Evolution of Communities* (Cody, M. L. and Diamond, J. M. eds), Harvard University Press, Cambridge, pp. 196—213.

[224] MacArthur, R. H. (1969), Patterns of communities in the tropics. *Biol. J. Linn. Soc.*, **1**, 19—30.

[225] Mayr, E. (1963), *Animal Species and Evolution.* Harvard University Press, Cambridge.

[226] Rosenzweig, M. L. and Taylor, J. A. (1980), Speciation and diversity in Ordovician invertebrates: filling niches quickly and carefully. *Oikos*, **35**, 236—43.

[227] Patrick, R. (1975), Structure of stream communities, in *Ecology and Evolution of Communities* (Cody, M. L. and Diamond, J. M. eds) Harvard University Press, Cambridge, pp. 445—59.

[228] Richerson, P. J. and Lum, K. (1980), Patterns of plant species diversity in California: relation to weather and topography. *Am. Nat.*, **116**, 504—36.
An immense library study examining the factors affecting plant diversity and relating the results to several hypotheses on the regulation of species diversity.

[229] MacArthur, J. W. (1975), Environmental fluctuations and species diversity, in *Ecology and Evolution of Communities* (Cody, M. L. and Diamond, J. M. eds.), Harvard University Press, Cambridge, pp. 74—80.

[230] Sanders, H. L. (1968), Marine benthic diversity: a comparative study. *Am. Nat.*, **102**, 243—82.

[231] Slobodkin, L. B. and Sanders, H. L. (1969), On the contribution of environmental predictability to species diversity. *Brookhaven Symp. Biol.*, **22**, 82—95.

[232] Abele, L. and Walters, K. (1979), The Stability—Time hypothesis: reevaluation of the data. *Am. Nat.*, **114**, 559—68.

[233] Fox, J. F. (1981), Intermediate levels of soil disturbance maximise alpine plant diversity. *Nature*, **293**, 564—65.

[234] Kummerer, R. W. and Allen, T. F. H. (1982), The role of disturbance in the pattern of a riparian bryophte community. *Am. Midl. Nat.*, **107**, 370—83.

[235] Hutchinson, G. E. (1961), The paradox of the plankton. *Am. Nat.*, **95**, 137—45.

[236] Herbert, P. D. N. (1977), Niche overlap among species in the *Daphnia carinata* complex. *J. Anim. Ecol.*, **46**, 399—410.

[237] Ranta, E. and Vespäläinen, K. (1981), Why are there so many species? Spatio-temporal heterogeneity and northern bumble-bee communities. *Oikos*, **36**, 28—34.

[238] Hassell, M. P. (1980), Some consequences of habitat heterogeneity for population dynamics. *Oikos*, **35**, 150—60.

[239] Fitter, A. H. (1982), Influence of soil heterogeneity on the coexistence of grassland species. *J. Ecol.*, **76**, 139—48.

[240] Hendrix, S. D. (1980), An evolutionary and ecological perspective of the insect fauna of ferns. *Am. Nat.*, **115**, 171—96.

[241] Moran, V. C. and Southwood, T. R. E. (1982), The guild composition of arthropod communities in trees. *J. Anim. Ecol.*, **51**, 289—306.
A tremendously detailed analysis of a complex insect assemblage.

[242] Boomsma, J. and Van Loon, A. (1982), Structure and diversity of ant communities in successive coastal dune valleys. *J. Anim. Ecol.*, **51**, 957—74.

[243] Woodin, S. A. (1981), Disturbance and community structure in a shallow water sand flat. *Ecology*, **62**, 1052—66.

[244] Schlosser, I. J. (1982), Fish community structure and function along two habitat gradients in a headwater stream. *Ecol. Monogr.*, **52**, 395—414.

[245] Augsburger, C. (1983), Offspring recruitment around tropical trees: changes in cohort distance with time. *Oikos*, **40**, 189—96.

[246] May, R. M. (1974), On the theory of niche overlap. *Theor. Pop. Biol.*, **5**, 297—332.
An involved theoretical analysis of niche overlap.

[247] Davidson, D. W. (1980), Some consequences of diffuse competition in a desert ant community. *Am. Nat.*, **116**, 92—105.

[248] Southwood, T. R. E. (1961), The number of species of insect associated with various trees. *J. Anim. Ecol.*, **30**, 1—8.

[249] Strong, D. R. (1974), The insects of British trees; community equilibration in ecological time. *Ann. Missouri Bot. Gard.*, **61**, 692—701.

[250] Southwood, T. R. E., Moran, V. C. and Kennedy, L. E. J. (1982), The richness, abundance and biomass of the arthropod communities on trees. *J. Anim. Ecol.*, **51**, 635—50.

[251] Birks, H. J. B. (1980), British trees and insects: a test of the time hypothesis over the last 13,000 years. *Am. Nat.*, **115**, 600—5.

[252] Pielou, E. (1975), *Ecological Diversity*. Wiley, New York.

[253] Fisher, R. A., Corbet, A. S. and Williams, C. B. (1943), The relation between the number of species and the number of individuals in a random sample of an animal population. *J. Anim. Ecol.*, **12**, 42—58.

[254] Preston, F. (1948), The commonness and rarity of species. *Ecology*, **29**, 254—83.

[255] Preston, F. (1962), The canonical distribution of commonness and rarity. *Ecology*, **43**, 410—32.

[256] Tepedino, V. J. and Stanton, N. L. (1981), Diversity and competition in bee—plant communities on short-grass prairie. *Oikos*, **36**, 35—44.

[257] Ugland, K. I. and Gray, J. S. (1982), Lognormal distributions and the concept of community equilibrium. *Oikos*, **39**, 171—78.

[258] MacArthur, R. H. (1957), On the relative abundance of bird species. *Proc. Nat. Acad. Sci.*, **43**, 293—95.

[259] King, C. E. (1964), Relative abundance of species and MacArthur's model. *Ecology*, **45**, 716—27.
A useful review on the adequacy of broken stick model in describing the relative abundance of various types of organism.

[260] DeVita, J. (1979), Niche separation and the Broken Stick model. *Am. Nat.*, **114**, 171—78.

[261] Pielou, E. and Arnasson, A. (1966), Correction to one of MacArthur's species abundance functions. *Science*, **151**, 592.

[262] MacArthur, R. H. (1966), Note on Mrs. Peilou's comments. *Ecology*, **47**, 1074.

[263] Cohen, J. E. (1968), Alternative derivations of a species—abundance relation. *Am. Nat.*, **102**, 165—72.

[264] Southwood, T. R. E., Brown, V. and Reader, P. (1979), The relationship of plant and insect diversities in succession. *Biol. J. Linn. Soc.*, **12**, 327—48.

[265] Lawton, J. H. (1982), Vacant niches and unsaturated communities: a comparison of bracken herbivores at sites on two continents. *J. Anim. Ecol.*, **51**, 573—95.
Presents more data suggesting that phytophagous insect assemblages are not structured to any major extent by interspecific competition, and presents a useful summary of species—area relationships for insects and host plants.

[266] Joseph Wright, S. and Biehl, C. (1982), Island biogeographic distributions; testing for random, regular and aggregated patterns of species occurrence. *Am. Nat.*, **119**, 345—57.

[267] Lawlor, L. R. (1980), Overlap, similarity and competition coefficients. *Ecology*, **61**, 245—51.

[268] Sale, P. F. (1974), Overlap in resource use and interspecific competition. *Oecologia*, **17**, 245—56.

[269] Pianka, E. R. (1981), Competition and niche theory, in *Theoretical Ecology. Principles and Applications* (May, R. M. ed.), Blackwell Scientific Publications, Oxford, pp. 167—96.
    A very readable account, coming down firmly on the side of competition as being the major organizing factor in communities.

[270] Cody, M. L. (1974), *Competition and the Structure of Bird Communities*. Princeton University Press, Princeton.

[271] Gladfelter, W. B., Ogden, J. C. and Gladfelter, E. H. (1980), Similarity and diversity among coral reef fish communities: A comparison between tropical western Atlantic (Virgin Islands) and tropical central Pacific (Marshall Islands) Patch reefs. *Ecology.*, **61**, 1156—1168.

[272] Casewell, H. (1976), Community structure: a neutral model analysis. *Ecol. Monogr.*, **46**, 327—54.

[273] May, R. M. (1979), The structure and dynamics of ecological communities, in *Population Dynamics* (Anderson, R. M., Turner, B. D. and Taylor, L. R. eds), Blackwell Scientific Publications, Oxford, pp. 385—408.
    Reviews the main themes that have emerged from theoretical and empirical studies of the dynamics of single populations and the mathematical models used to explore questions about the way multi-species systems are organized.

[274] Pimm, S. L. (1982), *Food Webs*. Chapman and Hall, Population and Community Biology series, London.
    The best introduction to the subject currently available.

[275] Pimm, S. L. and Lawton, J. H. (1977), Number of trophic levels in ecological communities. *Nature*, **268**, 329—31.
    Presents an alternative to the long-held energetics explanation to the length of food chains.

[276] Cohen, J. E. (1977), Ratio of prey to predators in community food webs. *Nature*, **270**, 165—67.

[277] Evans, F. C. and Murdoch, W. W. (1968), Taxonomic composition, trophic structure and seasonal occurrence in a grassland insect community. *J. Anim. Ecol.*, **37**, 259—73.

[278] Cole, B. J. (1980), Trophic structure of a grassland insect community. *Nature*, **288**, 77—8.

[279] Lakhani, K. H. (1982), Trophic structure of a grassland insect community. *Nature*, **299**, 375—76.

[280] Heatwole, H. and Levins, R. (1972), Trophic structure, stability and faunal change during recolonisation. *Ecology*, **53**, 531—34.

[281] Simberloff, D. S. and Wilson, E. O. (1969), Experimental zoogeography of islands: the colonisation of empty islands. *Ecology*, **50**, 278—95.

[282] Simberloff, D. S. (1976), Trophic structure determination and equilibrium in an arthropod community. *Ecology*, **57**, 395—98.

[283] Glasser, J. W. (1982), On the causes of temporal change in communities: modification of the biotic environment. *Am. Nat.*, **119**, 375—90.

[284] Wolda, H. (1978), Fluctuations in abundance of tropical insects. *Am. Nat.*, **112**, 1017—45.

[285] Zaret, T. M. (1982), The stability/diversity controversy: A test of hypotheses. *Ecology*, **63**, 721—31.

[286] May, R. M. (1972), Will a large complex system be stable? *Nature*, **238**, 413—14.

[287] Shorrocks, B., Atkinson, W. and Charlesworth, P. (1979), Competition on a divided and ephemeral resource, *J. Anim. Ecol.*, **48**, 899—908.

[288] McNaughton, S. J. (1978), Stability and diversity of ecological communities. *Nature*, **274**, 251—53.

[289] Rejmanek, M. and Stary, P. (1979), Connectance in real biotic communities and critical values for stability of model ecosystems. *Nature*, **280**, 311—13.

[290] Yodzis, P. (1980), The connectance of real ecosystems. *Nature*, **284**, 544—45.

[291] Zaret, T. M. and Paine, R. T. (1973), Species introduction in a tropical lake. *Science*, **182**, 449—55.
A fascinating example of how an introduced predator can produce population changes in a wide range of trophic levels.

[292] Cole, B. J. (1983), Assembly of mangrove ant communities: patterns of geographical distribution. *J. Anim. Ecol.*, **52**, 339—48.

[293] Preston, F. W. (1960), Time and space and the variation of species. *Ecology*, **29**, 254—83.

[294] Price, J. H. (1980), Niche and community in the inshore benthos with emphasis on the macroalgae, in *The shore environment, vol. 2*, (Price, J. H., Irvine, D. C. and Farnham, W. eds), Systematics Association special volume no. 17. pp 487—564.
Usefully reviews definitions and development of the niche and community concepts and systemmatically defines a number of terms such as hyperspace, niche breath, overlap and species packing.

[295] Tilman, D. C. (1980), *Resource competition and community structure.* Princeton University Press, Princeton.
An interesting monograph which concentrates on relationships between availability of limiting resources, competition and the structure of natural and manipulated plant assemblages.

[296] Schoener, T. W. (1983), Field experiments on interspecific competition. *Am. Nat.*, **122**, 240—85.
A very important review article, synthesising the results of all field experiments on interspecific competition available at the time. Part of a whole volume devoted to community ecology.

[297] Connell, J. H. (1983), On the prevalence and relative importance

of interspecific competition: evidence from field experiments. *Am. Nat.*, **172**, 661—96.
Written as a conflicting review to that of Schoener's above, but still demonstrates that interspecific competition is an important factor in community structure.

[298] Aarssen, L. W. (1983), Ecological combining ability and competitive combining ability in plants: toward a general evolutionary theory of coexistence in systems of competition. *Am. Nat.*, **122**, 707—31.

[299] Fenchel, T., Frier, J. and Kolding, S. (1978), The evolution of competing species, in *Marine organisms, genetics, ecology and evolution* (Battaglia, B. and Beardmore, J. eds). Marine Sciences vol. 2, Plenum Press, New York. pp. 289—301.
Reviews examples of character displacement of congeneric marine species, based on experimental evidence.

[300] Bradshaw, A. D. (1969), An ecologist's viewpoint, in *Ecological aspects of the mineral nutrition of plants* (I. Rorison ed.), Blackwell Scientific Publications, Oxford, pp. 415—27.

[301] Roughgarden, J. (1983), Competition and theory in community ecology. *Am Nat.*, **122**, 583—601.
Attacks the critics and criticisms of the role of competition in community ecology on philosophical grounds.

[302] Hughes, R. N. (1980), Predation and community structure, in *The shore environment. Vol. 2, Ecosystems*, (Price, J., Irvine, D. and Farnham, W. eds), Syst. Assoc. special volume, 17, Academic Press, London, pp. 699—728.

[303] Hunter, R. D. and Russell-Hunter, W. D. (1983), Bioenergetic and community changes in intertidal aufwuchs grazed by *Littorina littorea. Ecol.*, **64**, 761—69.

[304] Peterson, C. (1979), Predation, competitive exclusion and diversity in the soft sediment benthic communities of esturaries and lagoons, in *Ecological processes in coastal and marine systems* (Livingstone, R. J. ed.), Marine Sciences 10, Plenum Press, New York, pp. 233—63.

[305] Lubchenco, J. (1978), Plant species diversity in a marine intertidal community: importance of herbivore food preference and algal competitive abilities. *Am Nat.*, **112**, 23—39.

[306] Moreno, C. A., Sutherland, J. P. and Jara, H. F. (1984), Man as a predator in the intertidal zone of southern Chile. *Oikos*, **42**, 155—60.

[307] Williamson, M. (1983), The land bird community of Stockholm: ordination and turnover. *Oikos*, **41**, 378—84.

[308] Schoener, T. W. (1983), Rate of species turnover decreases from lower to higher organisms: a review of the data. *Oikos*, **41**, 372—77.
Part of a whole volume devoted to island biogeography.

[309] Connor, E. F. McCoy, E. D. and Cosby, B. J. (1983), Model discrimination and expected slope values in species-area studies. *Am. Nat.*, **122**, 789—96.

[310] Abbott, I. (1983), The meaning of Z in species/area regressions and the study of species turnover in island biogeography. *Oikos*, 41, 385—99.

[311] Abele, L. G. (1979), The community structure of coral-associated decapod crustaceans in variable environments, in *Ecological processes in coastal and marine systems* (Livingston, R. J. ed.), Marine Science 10, Plenum Press, New York, pp. 265—87.

[312] Rey, J. R. (1981), Ecological biogeography of arthropods on *Spartina alterniflora* islands in N.W. Florida. *Ecol. mong.*, 51, 237—65.
An experimental study of recolonisation of islands following defaunation.

[313] Ahlén, I. (1983), The bat fauna of some isolated islands in Scandinavia. *Oikos*, 41, 352—58.

[314] Blaustein, A. R., Kuris, A. M. and Alio, J. J. (1983), Pest and parasite species-richness problems. *Am. Nat.*, 122, 256—66.

[315] McCoy, E. D. and Rey, J. R. (1983), Area-related species richness: the uses of ecological and paleontological data. *Am. Nat.*, 122, 567—69.

[316] Southwood, T. R. E. and Kennedy, C. E. J. (1983), Trees as islands. *Oikos*, 141, 359—71.

[317] Haila, Y. (1983), Land birds on northern islands: a sampling metaphor for insular colonisation. *Oikos*, 41, 334—51.

[318] Bush, A. and Holmes, J. (1983), Niche separation and the broken stick model: use with multiple assemblages. *Am. Nat.*, 122, 849—55.

[319] Travis, J. and Ricklefs, R. E. (1983), A morphological comparison of island and mainland assemblages of neotropical birds. *Oikos*, 41, 434—41.

[320] Connor, E. F. and Simberloff, D. (1983), Interspecific competition and species occurance patterns on islands: null models and the evaluation of evidence. *Oikos*, 41, 455—65.
Reviews the types and supports the use and value of null models.

[321] Case, T. J. (1983), Niche overlap and assembly of island lizard communities. *Oikos*, 41, 427—33.

[322] Simberloff, D. (1983), Competition theory, hypothesis-testing, and other community ecological buzzwords. *Am Nat.*, 122, 626—35.

[323] Quinn, J. F. and Dunham, A. E. (1983), On hypothesis testing in ecology and evolution. *Am Nat.*, 122, 602—17.
Disputes the usefulness and logical primacy of 'null' models in descriptions of species distributions on islands.

[324] Briand, F. and Cohen, J. (1984), Community food webs have scale invariant structure. *Nature*, 307, 264—67.

[325] Pimm, S. L. (1984), The complexity and stability of ecosystems. *Nature*, 307, 321—26.

[326] King, A. W. and Pimm, S. L. (1983), Complexity, diversity and stability: a reconciliation of theoretical and empirical results. *Am. Nat.*, 122, 229—39.

Index